Piezoelectric MEMS

Piezoelectric MEMS

Special Issue Editors

Ulrich Schmid
Michael Schneider

MDPI • Basel • Beijing • Wuhan • Barcelona • Belgrade

MDPI

Special Issue Editors

Ulrich Schmid
Institute of Sensor and Actuator Systems
Austria

Michael Schneider
Institute of Sensor and Actuator Systems
Austria

Editorial Office
MDPI
St. Alban-Anlage 66
Basel, Switzerland

This is a reprint of articles from the Special Issue published online in the open access journal *Micromachines* (ISSN 2072-666X) from 2017 to 2018 (available at: http://www.mdpi.com/journal/micromachines/special_issues/piezoelectric_mems)

For citation purposes, cite each article independently as indicated on the article page online and as indicated below:

LastName, A.A.; LastName, B.B.; LastName, C.C. Article Title. *Journal Name* **Year**, *Article Number, Page Range.*

ISBN 978-3-03897-005-7 (Pbk)
ISBN 978-3-03897-006-4 (PDF)

Cover image courtesy of Ulrich Schmid, Michael Schneider and Georg Pfusterschmied.

Contents

About the Special Issue Editors

Ulrich Schmid started studying physics and mathematics at the University of Kassel in 1992. He performed his diploma work at the research laboratories of the Daimler-Benz AG on silicon carbide (6H-SiC) microelectronic devices. In 1999, he joined the research laboratories of DaimlerChrysler AG (now Airbus Group) in Ottobrunn/Munich, Germany. In 2003, he received his Ph.D. degree from the Technical University (TU) Munich. From 2003 to 2008, he was postdoc at the Chair of Micromechanics, Microfluidics/Microactuators at Saarland University. Since October 2008, he has held the position of full professor for Microsystems Technology at the Institute of Sensor and Actuator Systems at TU Vienna. Ulrich Schmid has authored or co-authored more than 350 peer-reviewed contributions in scientific journals and conferences and holds more than 40 granted patents. His research interests are in functional materials for MEMS/NEMS devices, such as aluminium nitride or silicon carbide, and in the modelling, simulation and evaluation of these MEMS/NEMS devices for different application scenarios.

Michael Schneider studied physics at the Karlsruhe Institute of Technology 2003 2009. He performed his diploma work at the Forschungszentrum Karlsruhe on the measurement of Lorentz angles in highly irradiated silicon strip detectors for high energy collider applications, such as the large hadron collider at CERN. He finished his studies in 2009 and started his PhD thesis on the optimization of ultra-thin aluminum nitride films for actuation and sensing applications in micro electromechanical systems at the Department of Microsystems Technology at TU Vienna. He received his PhD in 2014 and is currently working as a postdoc on advanced materials such as silicon carbide and doped aluminum nitride, as well as MEMS devices based on piezoelectric thin films.

micromachines

MDPI

Editorial

Editorial for the Special Issue on Piezoelectric MEMS

Ulrich Schmid * and Michael Schneider *

Institute of Sensor and Actuator Systems, TU Wien, 1040 Vienna, Austria
* Correspondence: ulrich.e366.schmid@tuwien.ac.at (U.S.); michael.schneider@tuwien.ac.at (M.S.)

Received: 10 May 2018; Accepted: 10 May 2018; Published: 15 May 2018

Electromechanical transducers that utilize the piezoelectric effect have been increasingly used in micro-electromechanical systems (MEMS) either as substrates or as thin films. Piezoelectric transducers feature a linear voltage response, no snap-in behaviour, and can provide both attractive and repulsive forces. Such features remove the inherent physical limitations present in the commonly used electrostatic transducer approach while preserving its beneficial properties such as low-power operation. Furthermore, piezoelectric materials are suitable for both actuation and sensing purposes; in addition to their compact design, they enable pure electrical excitation as well as read-out of the transducer element. On the basis of these characteristics, the operation of piezoelectric transducers suits a large variety of different application scenarios, ranging from resonators in advanced acoustic devices in liquid environments to sensors in harsh environments. To uncover the full potential of piezoelectric MEMS, interdisciplinary research efforts in a variety of subjects are needed, including investigations of advanced piezoelectric materials with regard to the design of novel piezoelectric MEMS sensor and actuator devices as well as the integration of PiezoMEMS devices into full low-power systems.

This special issue covers contributions to the current state of this exciting field of research in the following topics:

1. Experimental and theoretical research on the deposition, properties, and actuation structures of piezoelectric materials such as aluminum nitride (AlN) and lead zirconate titanate (PZT) with a focus on the application in MEMS devices. An et al. [1] presented an explanation of the causes of hysteresis effects in piezoelectric ceramic actuators using micropolarization theory. By employing a control method based on a tripartite Prandtl-Ishlinskii (PI) model, An et al. could improve the tracking performance by more than 80%. In addition, Qin et al. proposed a modification of the PI model which allows for the identification of all parameters of the hysteresis model through one set of experimental data, without the need for additional curve fitting [2]. Sette et al. [3] developed fully transparent PZT thin film capacitors contacted via Al-doped zinc oxide (AZO), which can be utilized to add new functionalities to transparent surfaces, such as providing in-display actuation for haptic feedback in mobile devices.

2. Modelling and simulation of piezoelectric MEMS devices and systems. Yang et al. [4] proposed and simulated a novel AlN on silicon cantilever gyroscope based on inversely connected electrode stripes, which offers a theoretical sensitivity of 0.145 pm/°/s at a small device footprint. Wei et al. established [5] general analytical equations based on a kinematic analysis of compliant bridge mechanisms, which were then used to optimize a piezo-driven compliant bridge mechanism. Li et al. [6] presented the dynamic characteristics of piezoelectric micro jets by utilizing a direct coupling simulation approach, including the impact of inlet and viscous losses. In a second paper, Li et al. [7] analyzed the impact of fluid density and acoustic velocity on the micro jet performance. Chen et al. [8] simulated and experimentally validated a PZT-actuated, triple-finger gripper, which reached an output resolution of 145 nm/V at a maximum displacement range of 43.4 μm.

3. Piezoelectric MEMS resonators for measuring physical quantities such as mass, acceleration, yaw rate, as well as the pressure, viscosity, or density of liquids. Pfusterschmied et al. [9] demonstrated

that piezoelectric MEMS resonators with high quality factors in liquids can be used to monitor the change in grape must during wine fermentation, which is a direct quality indicator of the fermentation process. Yu et al. [10] presented a unique take on the MEMS gyroscope through use of the acoustic Sagnac effect, which measured the phase difference between two sound waves traveling in opposite directions in a circular MEMS structure actuated by PMUTs.

4. Acoustic devices, such as surface acoustic wave (SAW), bulk acoustic wave (BAW), or thin film bulk acoustic resonators (FBARs) as well as acoustic transducers, which use piezoelectric MEMS, such as microphones or loudspeakers. Udvardi et al. [11] proposed a low-volume, piezoMEMS-based spirally shaped acoustic receptor array. The device is small enough to be used in cochlear implants while maintaining a good low-frequency response with output voltages high enough for direct analog conversion. Mansoor et al. [12] presented a transduction system for extremely fast system dynamics, which can create stationary as well as traveling surface waves in a turbulent boundary layer.

5. Piezoelectric energy harvesting technologies. Xu et al. [13] presented a hybrid meso-scale energy harvesting device, which combines both piezoelectric and electromagnetic harvesting schemes; under certain conditions, this hybrid approach can provide wider bandwidths and higher output power for vibrational energy harvesters.

References

1. An, D.; Li, H.; Xu, Y.; Zhang, L. Compensation of Hysteresis on Piezoelectric Actuators Based on Tripartite PI Model. *Micromachines* **2018**, *9*, 44. [CrossRef]
2. Qin, Y.; Zhao, X.; Zhou, L. Modeling and Identification of the Rate-Dependent Hysteresis of Piezoelectric Actuator Using a Modified Prandtl-Ishlinskii Model. *Micromachines* **2017**, *8*, 114. [CrossRef]
3. Sette, D.; Girod, S.; Leturcq, R.; Glinsek, S.; Defay, E. Transparent Ferroelectric Capacitors on Glass. *Micromachines* **2017**, *8*, 313. [CrossRef]
4. Yang, J.; Si, C.; Yang, F.; Han, G.; Ning, J.; Yang, F.; Wang, X. Design and Simulation of A Novel Piezoelectric AlN-Si Cantilever Gyroscope. *Micromachines* **2018**, *9*, 81. [CrossRef]
5. Wei, H.; Shirinzadeh, B.; Li, W.; Clark, L.; Pinskier, J.; Wang, Y. Development of Piezo-Driven Compliant Bridge Mechanisms: General Analytical Equations and Optimization of Displacement Amplification. *Micromachines* **2017**, *8*, 238. [CrossRef]
6. Li, K.; Liu, J.-K.; Chen, W.-S.; Zhang, L. Influences of Excitation on Dynamic Characteristics of Piezoelectric Micro-Jets. *Micromachines* **2017**, *8*, 213. [CrossRef]
7. Li, K.; Liu, J.-K.; Chen, W.-S.; Zhang, L. Comparative Influences of Fluid and Shell on Modeled Ejection Performance of a Piezoelectric Micro-Jet. *Micromachines* **2017**, *8*, 21. [CrossRef]
8. Chen, T.; Wang, Y.; Yang, Z.; Liu, H.; Liu, J.; Sun, L. A PZT Actuated Triple-Finger Gripper for Multi-Target Micromanipulation. *Micromachines* **2017**, *8*, 33. [CrossRef]
9. Pfusterschmied, G.; Toledo, J.; Kucera, M.; Steindl, W.; Zemann, S.; Ruiz-Díez, V.; Schneider, M.; Bittner, A.; Sanchez-Rojas, J.; Schmid, U. Potential of Piezoelectric MEMS Resonators for Grape Must Fermentation Monitoring. *Micromachines* **2017**, *8*, 200. [CrossRef]
10. Yu, Y.; Luo, H.; Chen, B.; Tao, J.; Feng, Z.; Zhang, H.; Guo, W.; Zhang, D. MEMS Gyroscopes Based on Acoustic Sagnac Effect. *Micromachines* **2017**, *8*, 2. [CrossRef]
11. Udvardi, P.; Radó, J.; Straszner, A.; Ferencz, J.; Hajnal, Z.; Soleimani, S.; Schneider, M.; Schmid, U.; Révész, P.; Volk, J. Spiral-Shaped Piezoelectric MEMS Cantilever Array for Fully Implantable Hearing Systems. *Micromachines* **2017**, *8*, 311. [CrossRef]

12. Mansoor, M.; Köble, S.; Wong, T.; Woias, P.; Goldschmidtböing, F. Design, Characterization and Sensitivity Analysis of a Piezoelectric Ceramic/Metal Composite Transducer. *Micromachines* **2017**, *8*, 271. [CrossRef]
13. Xu, Z.; Shan, X.; Yang, H.; Wang, W.; Xie, T. Parametric Analysis and Experimental Verification of a Hybrid Vibration Energy Harvester Combining Piezoelectric and Electromagnetic Mechanisms. *Micromachines* **2017**, *8*, 189. [CrossRef]

micromachines

MDPI

Article

Compensation of Hysteresis on Piezoelectric Actuators Based on Tripartite PI Model

Dong An *, Haodong Li, Ying Xu and Lixiu Zhang *

College of Mechanical Engineering, Shenyang Jianzhu University, Hunnan East Road No.9,
Hunnan New District, Shenyang 110168, China; lihaodong@stu.sjzu.edu.cn (H.L.); yxu@sypi.com.cn (Y.X.)
* Correspondence: andong@sjzu.edu.cn (D.A.); huangli@mail.neu.edu.cn (L.Z.);
   Tel.: +86-024-2469-0088 (D.A.); Tel.: +86-024-2469-4412 (L.Z.)

Received: 7 December 2017; Accepted: 24 January 2018; Published: 26 January 2018

Abstract: Piezoelectric ceramic actuators have been widely used in nanopositioning applications owing to their fast response, high stiffness, and ability to generate large forces. However, the existence of nonlinearities such as hysteresis can greatly deteriorate the accuracy of the manipulation, even causing instability of the whole system. In this article, we have explained the causes of hysteresis based on the micropolarization theory and proposed a piezoelectric ceramic deformation speed law. For this, we analyzed the piezoelectric ceramic actuator deformation speed law based on the domain wall theory. Based on this analysis, a three-stage Prandtl–Ishlinskii (PI) model (hereafter referred to as tripartite PI model) was designed and implemented. According to the piezoelectric ceramic deformation speed law, this model makes separate local PI models in different parts of piezoelectric ceramics' hysteresis curve. The weighting values and threshold values of the tripartite PI model were obtained through a quadratic programming optimization algorithm. Compared to the classical PI model, the tripartite PI model can describe the asymmetry of hysteresis curves more accurately. A tripartite PI inverse controller, PI inverse controller, and Preisach inverse controller were used to compensate for the piezoelectric ceramic actuator in the experiment. The experimental results show that the inclusion of the PI inverse controller and the Preisach inverse controller improved the tracking performance of the tripartite PI inverse model by more than 80%.

Keywords: piezoelectric actuators; hysteresis nonlinearity; Prandtl–Ishlinskii (PI) model; hysteresis compensation; micropolarization

1. Introduction

In recent years, the rapid development of ultraprecision machining technology has led to higher positioning accuracy standards of the micropositioning platform driven by some functional materials such as piezoelectric ceramics. Piezoelectric ceramics actuators (PCAs) have been widely used in precision positioning applications, such as scanning and microscopic technologies [1,2], micromanipulators [3], atomic force microscopes [4–6], and ultraprecision machine tools [7,8]. This is because of their ability to achieve high precision and versatility to be implemented over a wide range of applications [9]. However, the existence of hysteresis in PCAs often limits the operation performance of the actuators. Therefore, it is highly desirable to compensate for the hysteresis so that the piezoelectric devices can have a virtually linear relationship, or one-to-one mapping between the control signal and the output displacement [10]. Figure 1 presents the relationship between the displacement and the voltage across a piezoelectric actuator. It can be seen from the figure that when the voltage is applied across the piezoelectric ceramic, the step-up displacement curve does not coincide with the step-down displacement curve, and the displacement does not return to zero after the applied voltage is reduced to zero. This phenomenon is called the piezoelectric ceramic hysteresis phenomenon. This means that the actuator output displacement depends not only on the input or applied voltage at the present time, but also on the input history [11]. The intrinsic nonlinear and multivalued hysteresis in the piezoelectric actuator has the potential to cause an inaccuracy in, or even instability of, its applied system. The maximum error resulting from the hysteresis can be as much as 10–15% of the path covered [12]. It is obvious from the above analysis that the available approaches for the identification of piezoelectric-actuated stages containing the hysteresis and linear dynamics are still an open problem [13]. Therefore, establishing a precise control model of piezoelectric ceramics, and (based on this model) controlling the hysteresis nonlinearity of piezoelectric ceramics so as to improve the control precision of piezoelectric ceramics, has become a hot issue discussed by many scholars, globally [14,15].

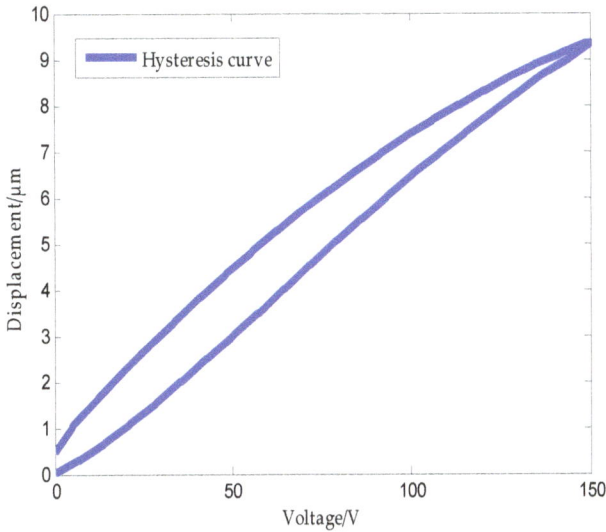

Figure 1. Hysteresis characteristic of a piezoelectric ceramic.

As increasingly more researchers focus on PCAs, there have been numerous attempts to use models to compensate for hysteresis. These are broadly divisible into two, namely, physics-based models (white/gray box) and phenomenological models (black box) [16]. The physics-based models are derived from the physical means of the hysteresis and can be strictly verified. The physical model refers to a scientific concept that is abstracted out from a large number of experiments for the convenience of research, excluding secondary factors and highlighting the main factors. One of the advantages of physics-based models is their clear physical meaning. However, due to the complicated form, physics-based models are not commonly used in the control of PCAs [17]. The commonly used hysteresis models based on the hysteresis nonlinearity of piezoelectric ceramics are the Jiles–Atherton (J–A) model [18] and the Maxwell model [19]. Malczyk et al. proposed an extension of the J–A model, as the J–A magnetic hysteresis model, to describe the hysteresis curve narrowing phenomenon in ferrite ZnMn material. Their new model permits the inclusion of a wide variety of additional effects observed in ferromagnetic materials without invalidating the well known and broadly used J–A model parameters. Experiments prove the feasibility of this method [20]. Liu et al. presented a Maxwell model to describe the hysteresis in a piezoelectric actuator. They studied the effect of the number of elements and presented both the forward and inverse algorithms. Further, they used the inverse Maxwell model and obtained almost linear performances of the hysteresis compensation. The results of their experiment validate the effectiveness of the proposed algorithm and showed a reduction in hysteresis nonlinearity from 13.8 to 0.4% [21]. Phenomenon-based models are the ones in which researchers generalize and summarize input and output data and the phenomena of practical experiments, utilizing mathematical methods to directly build a mathematical model to satisfy the experiment rule regardless of the physical meaning, such as the Preisach model [22], the Prandtle–Ishlinskii (PI) model [23], the Duhem model [24], and the Bouc–Wen model. Song et al. proposed a novel modified Preisach model to identify and simulate the hysteresis phenomenon observed in a piezoelectric stack actuator. Their approach can handle a varying-frequency dependence by employing a time-derivative correction technique. Parameter estimation and model verification demonstrated high accuracy of the derived model, keeping the deviation in a low percentage range (about 2–3%) [16]. Lin et al. reformulated the Bouc–Wen model, the Dahl model and the Duhem model as a generalized Duhem model to compare the performances of variant hysteresis models with respect to the tracking reference. Since the Duhem model includes both the electrical and mechanical domains, it has a smaller modeling error compared to the other two hysteresis models. Finally, a real-time experiment confirmed the feasibility of their proposed method [25]. Wang et al. proposed a novel modified Bouc–Wen (MBW) model to describe the asymmetric hysteresis of a piezoelectric actuator. They used a polynomial-based non-lag component to realize the asymmetric hysteresis property. The results demonstrate that their model is superior to its competitors' models in describing the asymmetric hysteresis of a piezoelectric actuator [26]. However, the lack of a physical meaning makes the above-mentioned model difficult to understand. Simultaneously, none of the abovementioned models reveal the cause of hysteresis from a microscopic point of view, thus, modeling errors in these modeling methods are inevitable.

In this study, we first analyze the causes of the hysteresis based on the micropolarization mechanism. Then, by observing the hysteresis curve of piezoelectric ceramic and establishing the deformation speed law of piezoelectric ceramics, we explain the deformation rate of piezoelectric ceramics at different stages, making use of the nucleation rate of microscopic domain evolution. After that, according to the proposed piezoelectric ceramic deformation speed law, we split and then recombined the play operator, and the improved PI model is proposed. Finally, the improved PI model is compared with the traditional PI model and Preisach model. The experimental results show that the accuracy of the improved PI model is increased by more than 80% as compared to the traditional PI model and Preisach model.

This paper is organized as follows: Section 2 describes hysteresis based on the microscopic polarization mechanism and domain wall theory, and reveals the cause of hysteresis from the microscopic point of view. A novel piezoelectric ceramic deformation speed law is proposed, and its analysis presented in Section 3. Section 4 presents the proposed tripartite PI model based on the deformation rate law of piezoelectric ceramics. A contrast experiment with traditional PI model and Preisach model is presented in Section 5. Finally, Section 6 provides a summary of discussion and future works.

2. Causes of Hysteresis

2.1. Micromechanism

The piezoelectric ceramics are obtained from ferroelectric ceramics after the polarization treatment, and thus, the property of piezoelectric ceramics is consistent with those of ferroelectric piezoelectric dielectric materials. Under the influence of an electric field, they have electrostriction effect, inverse piezoelectric effect, and ferroelectric effect [27].

The electrostriction effect is caused by dielectric polarization. In the presence of an electric field, dielectric molecules get polarized, thereby generating dielectric stress and the corresponding deformation. However, due to the strong mutual attraction between the nucleus and the electrons, the applied electric field is not sufficient to destroy the dielectric property; moreover, compared with the piezoelectric effect, the electrostrictive coefficient is several orders of magnitude smaller than the piezoelectric coefficient; hence, the electrostriction effect is extremely weak in the macro performance [28], and therefore, the output displacement of the piezoelectric ceramic can be ignored.

Curie brothers while studying quartz crystals in 1880 detected crystal deformation [29]. Under the effect of an external force, the surface of the crystal will have polarized charges when a mechanical force is applied. This appearance of electrical polarization is called direct piezoelectric effect, as shown in Figure 2a. On the contrary, if an electric field is applied to the piezoelectric crystal, the crystal not only produces polarization, but also produces deformation. This phenomenon of deformation caused by the electric field is called the inverse piezoelectric effect, as shown in Figure 2b. Piezoelectric ceramic output displacement feature is due to the inverse piezoelectric effect. In general, inverse piezoelectric effect can be expressed as

$$S = dE \tag{1}$$

where S is the strain due to the electric field, d is the piezoelectric constant, and E is the applied electric field strength. The inverse piezoelectric effect can be deduced from the above equation, is linear, and there are no hysteresis characteristics.

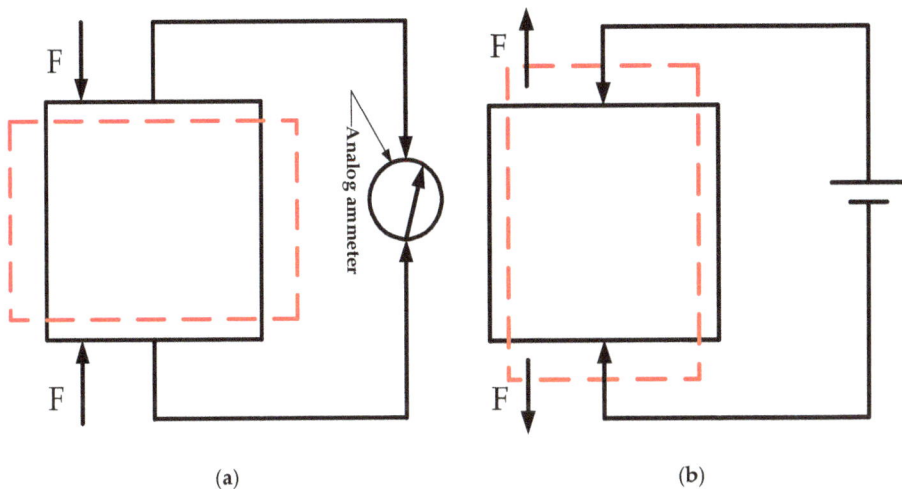

(a)   (b)

Figure 2. Piezoelectric effect diagram (Red dashed lines indicate after deformation): (**a**) Direct piezoelectric effect diagram; (**b**) Inverse piezoelectric effect diagram. The black rectangle represents the original shape of the piezoelectric ceramic block, and the red dashed rectangle represents the deformed shape.

Piezoelectric ceramic is a kind of ferroelectric material. Inside the piezoelectric ceramic, in the presence of an external force, the intrinsic dipole moments of the unit cell are arranged neatly in the same direction and cause the piezoelectric ceramic crystal to be in a highly polarized state. Spontaneous polarization in ferroelectric materials always splits into a series of small regions with different polarization directions, so that the electric fields established by spontaneous polarization with the external space offset each other. Therefore, the entire single crystal is nonelectrical. These small areas with the same direction of spontaneous polarization are called domains. There are usually four directions inside a piezoelectric ceramic transducer: the 71° domain, the 90° domain (as shown in Figure 3), the 109° domain, and the 180° domain. It should be noted that, for the crystal strain, only a non-180° domain steering contributes to the displacement of the PCAs, while a 180° domain steering has no effect on the volume effect [30]. Spontaneous polarization of the domain will reorient under the influence of an external electric field. This phenomenon of reorientation of the spontaneous polarization in a piezoelectric ceramic in the presence of an external electric field is known as the ferroelectric effect.

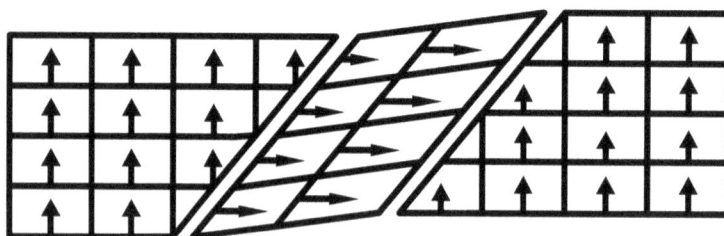

Figure 3. Piezoelectric crystal domain diagram.

Therefore, we define the micropolarization mechanism of a piezoelectric ceramic as if the direction of the applied electric field in a piezoelectric ceramic is the same as the polarization direction. Then, the domain inside the piezoelectric ceramic will have a certain degree of steering and elongation and the boundary of the domain will also produce elongation deformation. Therefore, the piezoelectric ceramic will have an elongation deformation along the polarization direction (as shown in Figure 4).

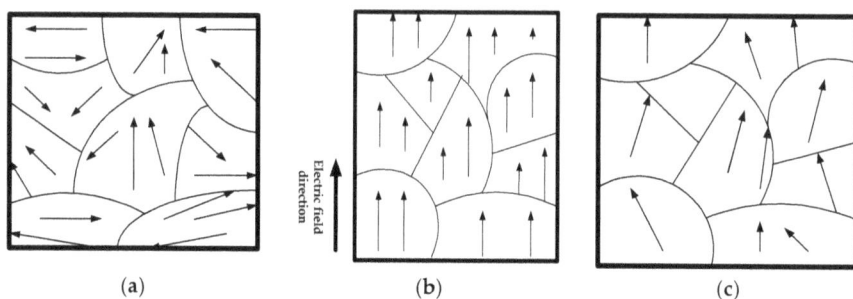

(a)   (b)   (c)

Figure 4. Schematic diagram of the spontaneous polarization alignment: (**a**) before; (**b**) during; (**c**) after presence of an electric field.

2.2. Analysis of the Causes of Hysteresis

When the applied electric field strength exceeds a certain critical field strength (the field strength that begins to turn the electric domain), the strain of the piezoelectric ceramic (except for the inverse piezoelectric effect) occurs, thus steering the non-180° domain (which is not completely reversible) and gradually starts to dominate. When the field strength is on the decline, some non-180° domains cannot be restored to the same level as at the time of increasing field strength.

In this study, we assume that $N1$ is the number of unit cells making non-180° domain turns in the piezoelectric ceramic when the field strength is increased and $N2$ is the number of unit cells making non-180° domain turns in the piezoelectric ceramic when the field strength is reduced. From the above analysis, we can conclude that $N1 > N2$; this partially irreversible non-180° domain causes the hysteresis in the displacement of the PCAs. Furthermore, the greater the field strength, the more irreversible the non-180° domain, and greater the hysteresis displacement of the PCAs.

3. Piezoelectric Ceramic Deformation Speed Law

3.1. Derivation of Deformation Speed Law

We utilized the Renishaw XL-80 (Renishaw plc, Gloucestershire, UK) laser interferometer (shown in Figure 5) to measure the deformation rate of piezoelectric ceramics for voltages ranging from 0 V to 150 V. Figure 6 shows the deformation rates of the PCAs for an applied triangle wave voltage of 150 V driven by different frequencies. Figure 7 shows the deformation rates of the PCAs for an applied triangular wave, a sine wave, and a manually added voltage.

As can be seen from Figure 6, although the triangular wave voltage frequency is different, the three sub-plans followed the same law: the deformation rate in the lift stage is below the timeline (which is the elongation rate), showing the trend of first increasing and then decreasing with time, during the boost period. Deformation rate change is not monotonic and the maximum value is taken as shown by the arrow in the figure. The return deformation rate is above the timeline (which is the contraction rate), showing an increasing trend over time: in the voltage reduction phase, deformation rate increases monotonically, with the maximum appearing at the end of voltage reduction phase. Figure 7 shows that although the applied voltage wave forms are different, the three sub-plans follow the above law.

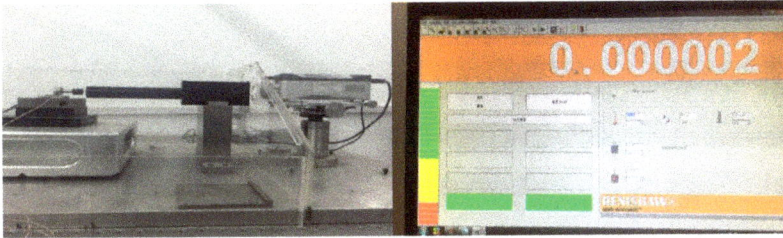

Figure 5. XL-80 laser interferometer ((Renishaw plc, Gloucestershire, UK)).

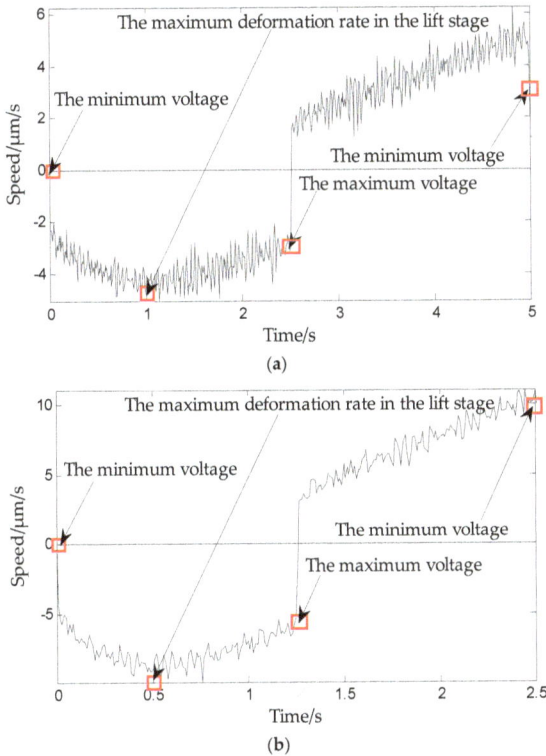

(a)

(b)

Figure 6. *Cont.*

(c)

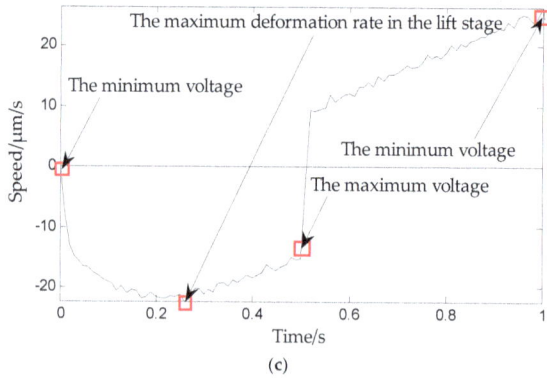

Figure 6. Deformation rate of piezoelectric actuators at an applied triangle wave voltage of 150 V and a frequency of (**a**) 0.2 Hz; (**b**) 0.4 Hz; (**c**) 1 Hz. Below the timeline, the voltage is loaded from the minimum voltage (0 V) to the maximum voltage (150 V). Above the timeline, the voltage drops from the maximum voltage (150 V) to the minimum voltage (0 V).

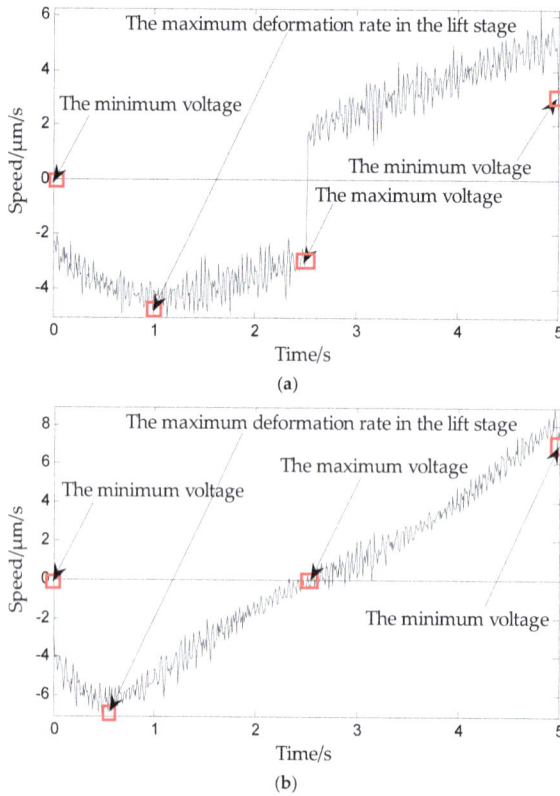

(a)

(b)

Figure 7. *Cont.*

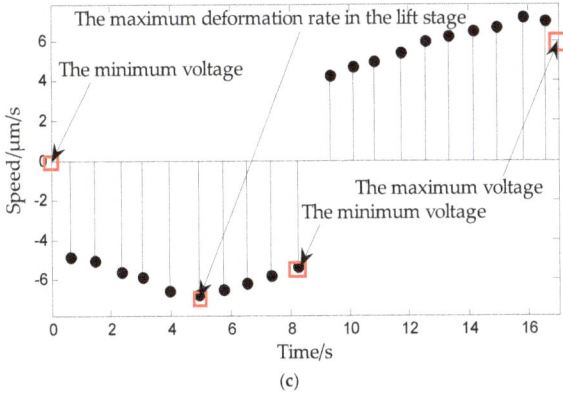

(c)

Figure 7. Deformation rate of piezoelectric actuators for an applied voltage of 150 V with frequency 1 Hz in (**a**) triangular wave form; (**b**) sign-wave form, $u = 150(\sin \pi t/5)$ positive half cycle; (**c**) Manually added, 0 V–150 V–0 V, at steps of 15 V. Below the timeline, the voltage is loaded from the minimum voltage (0 V) to the maximum voltage (150 V). Above the timeline, the voltage drops from the maximum voltage (150 V) to the minimum voltage (0 V).

Therefore, we propose the deformation rate law of piezoelectric ceramics: In the phase of voltage increase, the deformation rate of piezoelectric ceramics first increases and then decreases, and there is an inflection point voltage. During the voltage drop phase, the deformation rate of piezoelectric ceramics increases monotonically without the inflection point.

3.2. Analysis of Deformation Speed Law

In 2007, Rabe et al. proposed that when an electric domain turns under the influence of an electric field, the entire domain is not oriented like a dipole. Instead, the following four stages occur: new domain nucleation, vertical growth of new domain, horizontal expansion of the new domain, and new domain merger [31].

In this study, we analyze the above law based on the nucleation rate of microscopic domain evolution. As already mentioned in Section 2.1, piezoelectric ceramic deformation is due to the internal electric domain steering. Experiments of predecessors have confirmed that the physical mechanism of electrical domain inversion is the nucleation process and the nucleation rate of the domain is a function of the applied electric field [32]. Hence, through the change of nucleation rate, one can obtain the domain inversion volume change rate. Merz et al. obtained the relationship between the new domain nucleation rate and the applied load through experiments, generally conducted in the lower electric field range ($E = 0.1$ kV/cm–1.0 kV/cm). They found the nucleation rate in line with the exponential relationship [33]

$$n_1 = k_1 \exp\left(\frac{-\delta}{E}\right) \tag{2}$$

In the higher electric field range ($E > 1.0$ kV/cm), the nucleation rate conforms to the power function

$$n_2 = k_2 E^{1.4} \tag{3}$$

where n_1, n_2 represent the numbers of nucleations per unit time per unit area, δ is the activation of the electric field, and k_1, k_2 are constants.

We assume that the electric field changes uniformly. Taking saturated electric field as $2E_c$, the total number of domains contained in the piezoelectric ceramic is

$$N = \int_0^{E_c} k_1 \exp\left(\frac{-\delta}{E}\right) dE + \int_{E_c}^{2E_c} k_2 E^{1.4} dE \tag{4}$$

In a certain electric field, flipped domains are only related to the applied electric field. Therefore, the deformation rate of piezoelectric ceramics at low electric field strengths can be expressed as

$$\alpha_1 = \frac{\int_0^{E} k_1 \exp\left(\frac{-\delta}{E}\right) dE}{\int_0^{E_c} k_1 \exp\left(\frac{-\delta}{E}\right) dE + \int_{E_c}^{2E_c} k_2 E^{1.4} dE} \tag{5}$$

At high electric field strengths, the deformation rate can be expressed as

$$\alpha_2 = \frac{\int_0^{E_c} k_1 \exp\left(\frac{-\delta}{E}\right) dE + \int_{E_c}^{E} k_2 E^{1.4} dE}{\int_0^{E_c} k_1 \exp\left(\frac{-\delta}{E}\right) dE + \int_{E_c}^{2E_c} k_2 E^{1.4} dE} \tag{6}$$

We define E_0, E_1, E_2, E_3 ... , E_{n-1}, E_n as the field strength points at equal intervals on the axis of coordinates, while the distance between the two adjacent pressure points is defined as $h_i = E_i - E_{i-1}$. Assuming that E_c is an inflection point electric field (can be considered as a constant), and using the geometric meaning of definite integral, we can get

$$\alpha_1 = \frac{\sum_{i=1}^{c} h_i \exp\left(\frac{-\delta}{E}\right)}{M} \tag{7}$$

In the formula $M = \int_0^{E_c} k_1 \exp\left(\frac{-\delta}{E}\right) dE + \int_{E_c}^{2E_c} k_2 E^{1.4} dE$, since E_c is a constant, M can also be considered as a constant. Similarly,

$$\alpha_2 = \frac{m + \sum_{i=c}^{n} h_i k_2 E^{1.4}}{M} \tag{8}$$

where $m = \int_0^{E_c} k_1 \exp\left(\frac{-\delta}{E}\right) dE$ is also a constant.

As can be seen from Equations (7) and (8), α_1 and α_2 are electric field functions. However, the exponential function grows much faster than the power function. Therefore, the growth rate of α_1 is obviously greater than the growth rate of α_2, which means that the deformation rate of the piezoelectric ceramics in the range $0 \sim E_c$ is greater than the deformation rate of piezoelectric ceramics in the range $E_c \sim 2E_c$. In the piezoelectric deformation curve, the deformation speed of the voltage rise phase first increases and then decreases, and there is an inflection point of deformation rate. Combining this with the applied voltage period, we can determine the piezoelectric ceramic hysteresis curve inflection point voltage.

Based on this basic fact, this study proposes a tripartite PI model—a modeling method for the hysteresis characteristics curves of piezoelectric ceramics.

4. Hysteresis Modeling

Whether in scanning tunneling microscopes, atomic force microscopes, or other precision positioning systems, quick and accurate positioning of the probe is desired. However, due to hysteresis, it is difficult for the probes to locate the correct position quickly and accurately. Various ways to reduce errors and improve the positioning accuracy are currently in practice. In this paper, we discuss the use of modeling methods, as shown in Figure 8.

From the piezoelectric ceramic hysteresis curve shown in Figure 8, it can be seen that every time the voltage rises and reduces, a hysteresis error is introduced; however, the maximum error occurs in the main hysteresis loop (the distance from A to B in Figure 8). Therefore, in this study, we only model the main hysteresis loop of the hysteresis curve. As a next step, we will conduct a study on the remaining displacements of the hysteresis curve.

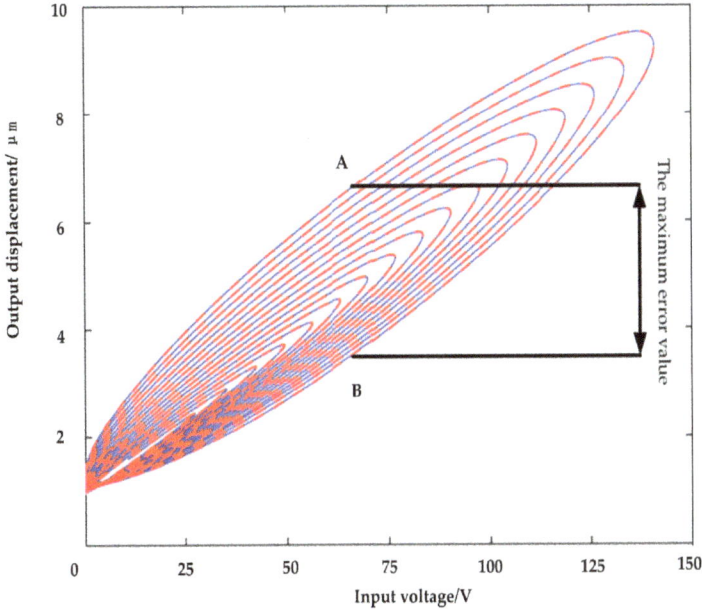

Figure 8. Piezoelectric ceramic input-output hysteresis curve.

4.1. Play Operator and Prandtle–Ishlinskii Model

The PI hysteresis model can be thought of as consisting of a stack of hysteresis operators. The mathematical expression for the play hysteresis operator shown in Figure 9a is

$$y(k) = \max\{u(k) - r, \min[u(k) + r, y(k-1)]\} \tag{9}$$

where k is the input time, r is the threshold of the play operator, $u(k)$ is the input of the operator, and $y(k)$ is the output of the operator. The initial value of hysteresis operator is defined as

$$y(0) = \max\{u(0) - r, \min[u(0) + r, h_0]\} \tag{10}$$

If the piezoelectric actuator is started from the power off state, the value of h_0 is 0.

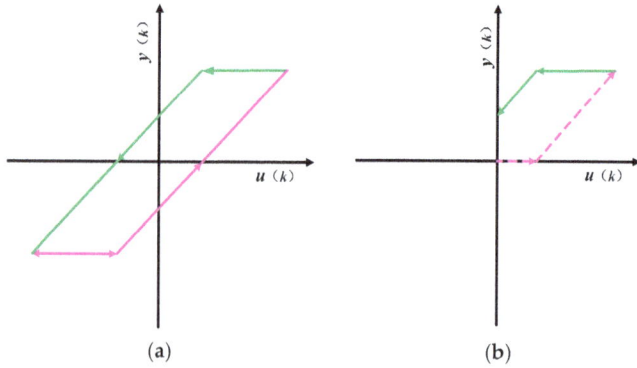

(a)   **(b)**

Figure 9. (**a**) Complete play operator; (**b**) play operator in practical application.

The PI hysteresis model was established in 1970 by the Russian mathematician Krasnoselskii, developed from the Preisach model and it was referred to as the PI model. It is formed by different play operators with different thresholds. The play operator is similar to the hysteresis curve in shape. Different play operators are multiplied by different weighting values and superimposed on one another to obtain the piezoelectric ceramic hysteresis PI model. The mathematical expression for the PI model is

$$Y(k) = \sum_{i=1}^{n} w_i \times y_i(k) = \sum_{i=1}^{n} w_i \times \max\{u(k) - r_i, \min[u(k) + r_i, y(k-1)]\} \tag{11}$$

where w_i is the weight of each hysteresis operator in the mathematical sense, n is the number of operators, $Y(k)$ is the output of the model at the moment k, and r_i is the threshold of the hysteresis operator. The vector form of Equation (11) is

$$Y(k) = \mathbf{w}^T \times \mathbf{y}(k) \tag{12}$$

where the threshold vector is $W = (w_1, \cdots, w_i, \cdots, w_n)^T$, the state vector of the operator at the moment k is $y(k) = (y_1(k), \cdots, y_i(k), \cdots, y_n(k))^T$, and the state vector of the operator at initial time is $y(0) = (y_1(0), \cdots, y_i(0), \cdots, y_n(0))^T$.

In the actual experiment, the step-down phase of the standard play operator is only partially present in the first quadrant, leading to its limited accuracy in describing the backhaul part of the hysteresis curve. The input voltage is always positive, which is increased from 0 V, therefore, in actual modeling, only a portion of the standard play operator is used. The input and output of the unilateral play operator are completely cuffed in the first quadrant, as shown in Figure 9b. The dotted and solid lines shown in Figure 9b are the output of the operator during increasing and decreasing voltage times. For $u(k) \leq r$, the output $y(k)$ always remains zero. For an input $r \leq u(k) \leq u_{\max}$, the operator output is $u(k) - r$. When the input voltage drops from peak $u_{\max}$ to $u_{\max} - 2r$, the operator output is $u_{\max} - r$. After this, the operator output $y(k)$ is $u(k) + r$, until the voltage drops to zero. The output of the operator whose threshold $r \geq 0.5u_{\max}$ does not have an output of $u(k) + r$.

4.2. Traditional PI Modeling and Inverse Model

We define $d(k)$ as the output displacement that corresponds to the input voltage of the piezoelectric ceramic at time k; the expression of the error signal $e(k)$ at time k is

$$e(k) = d(k) - Y(k) = d(k) - \mathbf{w}^T \times \mathbf{y}(k) = d(k) - \mathbf{y}^T(k) \times \mathbf{w} \tag{13}$$

The square of the error $e(k)$ is

$$e^2(k) = d^2(k) - 2d(k)\mathbf{y}^T(k)w + w^T\mathbf{y}(k)\mathbf{y}^T(k)w \tag{14}$$

In this study, the accuracy of modeling is measured by the addition of the squared errors $\sum_{k=1}^{n} e^2(k)$, where n is the number of sampling points. The weight vector w of the objective function is obtained from the quadratic programming algorithm, i.e.,

$$\begin{aligned} f(w) &= \sum_{k=1}^{n} e^2 = \sum_{k=1}^{n} [d(k) - \mathbf{y}^T(k)]^2 \\ &= \sum_{k=1}^{n} d^2(k) - 2\sum_{k=1}^{n} d(k)\mathbf{y}^T(k)w + \sum_{k=1}^{n} w^T\mathbf{y}(k)\mathbf{y}^T(k)w \end{aligned} \tag{15}$$

The cross-correlation function row vector R_{xd}^T and autocorrelation function matrix R_{xx} are defined as

$$R_{xd}^T = \sum_{k=1}^{n} d(k)\mathbf{y}^T(k) \tag{16}$$

$$R_{xx} = \sum_{k=1}^{n} \mathbf{y}(k)\mathbf{y}^T(k) \tag{17}$$

Equation (15) can now be expressed as

$$f(w) = \sum_{k=1}^{n} d^2(k) - 2R_{xd}^T w + wR_{xx}w^T \tag{18}$$

This indicates that $f(w)$ is a quadratic function of the weight coefficient vector w, which is an upwardly concave parabolic surface and a function with a unique minimum. The weight coefficient is adjusted so that $f(w)$ is the minimum, i.e., we find the minimum drop along the curved surface corresponding to the parabolic path. Here, we use the gradient descent method to find this minimum.

For Equation (18), we take a derivative with respect to the weight coefficient w, and we obtain the gradient of $f(w)$ as

$$\begin{aligned} \nabla(k) &= \nabla f(w) \\ &= -2R_{xd} + 2R_{xx}w \end{aligned} \tag{19}$$

Make $\nabla(k) = 0$, and the optimal weight coefficient vector can be obtained.

$$w = R_{xx}^{-1}R_{xd} \tag{20}$$

In order to ensure the positive definite of the quadratic matrix, the principle of threshold selection is $r_{max} < u_{max}$, $r_i = \frac{i}{n}\max|u(k)|$, $i = 0, 1, 2, \cdots, n-1$. Using a program written in MATLAB software (R2017b, MathWorks, Inc., Natick, MA, USA) for the operation, the parameters of the traditional PI model are obtained. They are tabulated in Table 1.

The triangular wave form shown in Figure 10 was applied to the piezoelectric ceramic. The experimental hysteresis curve of the piezoelectric ceramic and the curve obtained using the traditional PI model along with the error plot are shown in Figure 11. Error analysis using MATLAB software shows that the mean absolute error of the traditional PI model is $\delta_1 = \frac{1}{n}\sum_{k=1}^{n} |e_k| = 0.13947$ μm.

Table 1. Parameters of the Prandtl–Ishlinskii (PI) model. (*i* is the number of sampling point, r_i is the threshold of the play operator, w_i is the weight of each hysteresis operator in the mathematical sense).

i	r_i	w_i
1	0	0.0493
2	15	0.0298
3	30	0.0120
4	45	0.0090
5	60	0
6	75	0
7	90	0
8	105	0
9	120	0
10	135	0

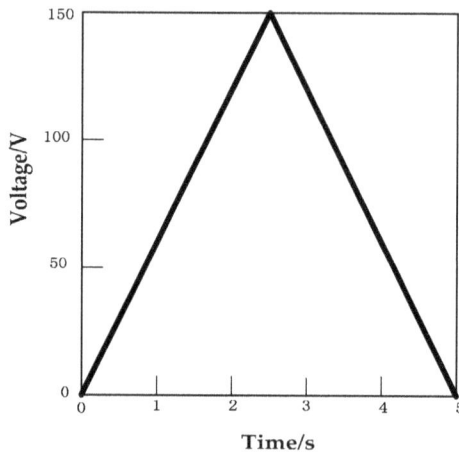

Figure 10. Input voltage of the piezoelectric ceramic actuator.

The inverse PI model is also a PI model. The threshold vector and the weight vector of the PI inverse model can be calculated using the relationship between the PI model and its inverse model. The PI model has an analytical inverse, $r'_i = \sum_{j=1}^{i} w_j (r_i - r_j)$, $i = 1, \cdots, n$; $w'_1 = \frac{1}{w_1}$,

$$w'_i = -w_i / \left[\left(\sum_{j=1}^{i} w_j \right) \left(\sum_{j=1}^{i-1} w_j \right) \right], \quad i = 2, \cdots, n; \quad u_i[0] = \sum_{j=1}^{i-1} w_j y_i[0] + \sum_{j=i}^{n} w_j y_j[0], \quad i = 2, \cdots, n.$$

Hence, the output expression of the PI inverse model at the time k is

$$U(k) = \sum_{i=1}^{n} w'_i \times u_i(k) = \sum_{i=1}^{n} w'_i \times \max\{y(k) - r'_i, \min[y(k) + r'_i, u(k-1)]\} \tag{21}$$

Figure 12 shows the experimental hysteresis curves of the piezoelectric ceramic and the ones obtained using the PI inverse model along with the error plot.

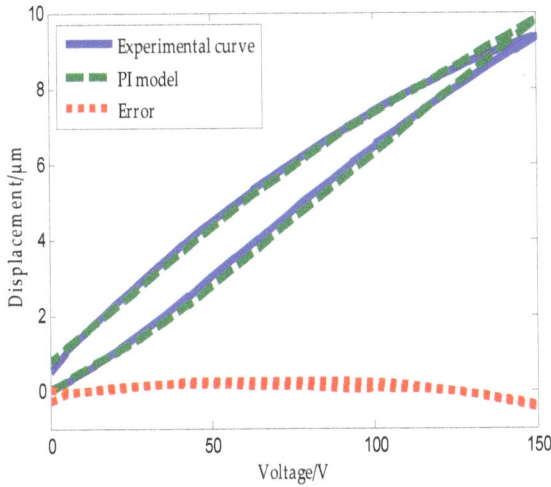

Figure 11. Hysteresis curve of piezoelectric ceramic actuators and PI model.

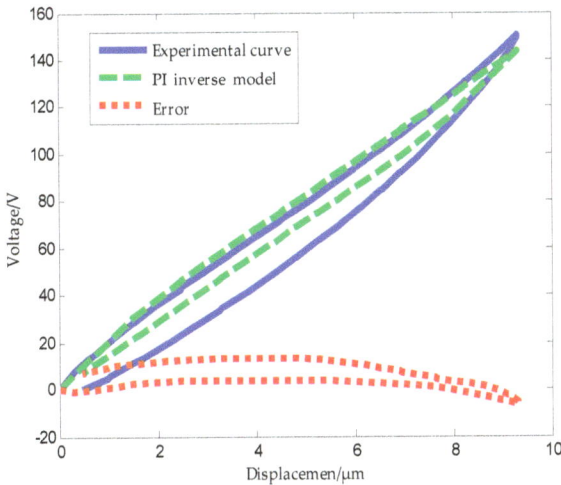

Figure 12. Hysteresis curve of PI inverse model.

Error analysis using MATLAB software shows that the mean absolute error of the traditional PI inverse model is $\delta_2 = \frac{1}{n}\sum_{k=1}^{n}|e_k| = 0.29435$ μm. From the above analysis, we can conclude that the traditional PI model and its inverse model exhibit large modeling errors.

4.3. Tripartite PI Model Based on the Deformation Rate of Piezoelectric Ceramics

In Section 3.1, we obtained the piezoelectric ceramic deformation speed law, and here, we propose a tripartite PI modeling method based on this law. As already mentioned in Section 4.1, the step-down phase of the standard play operator is only partially present in the first quadrant, therefore, the standard play operator has a limited description of the backhaul of the hysteresis curve. The operator used in the tripartite PI modeling method is a unilateral play operator. The input and output of the unilateral play operator are completely limited to the first quadrant, as shown in Figure 13.

The output expression of unilateral play operator is

$$\begin{cases} y(0) = \max\{u(0) - r, \min[u(0), 0]\} \\ y(k) = \max\{u(k) - r, \min[u(k), y(k-1)]\} \end{cases} \tag{22}$$

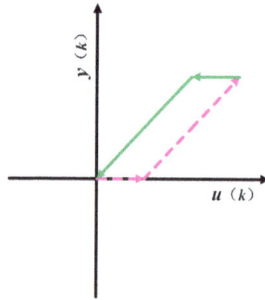

Figure 13. Schematic of one side play operator.

The dashed part is the boost part and the solid part is the buck part. The model is based on the theory presented in Section 3. A data collection experiment of the piezoelectric ceramic hysteresis curve under the triangle wave pressure provides support for modeling. The inflection point is captured based on Figures 6 and 7, which show the maximum deformation speed in the rising process.

Modeling steps:

(1) The selection of operators is based on the principles of concave-convex consistency, which means that in the hysteresis curve, the concave and convex parts of the curve correspond to the boost part and the depressurization part of the play operator, respectively.

(2) The rising curve rises from zero voltage to the inflection point voltage u_{if} (u_{if} refers to the voltage indicated by the arrows in Figures 6 and 7), i.e., when the deformation speed rises from 0 to the maximum. The relationship between the voltage and displacement is described by a single lateral play operator as shown in Figure 13 (the dotted portion).

(3) The rising curve rises from the inflection voltage u_{if} to maximum voltage u_{max} (u_{max} refers to the maximum point voltage applied to the piezoelectric ceramic during the whole rising cycle. It is 150 V here). Voltage–position relation in this part is described by a single lateral play operator as shown in Figure 13 (the solid line). One side play operators and hysteresis curves have a counter clock directivity. The reducing portion and rising process in the second part manifest the epirelief characteristic. The reducing portion of play operators point to the origin of coordinates while the second rising hysteresis curve deviates from it. Therefore, we need to model in reverse when we use play operators in the reducing part to describe the second rising process of the hysteresis curve.

(4) The retraced curve's relation that reduces from the maximum to zero voltage is described by a single lateral play operator as shown in Figure 13 (the solid line).

The application of operators in respective parts of the hysteresis curves during the model process are shown in Figure 14.

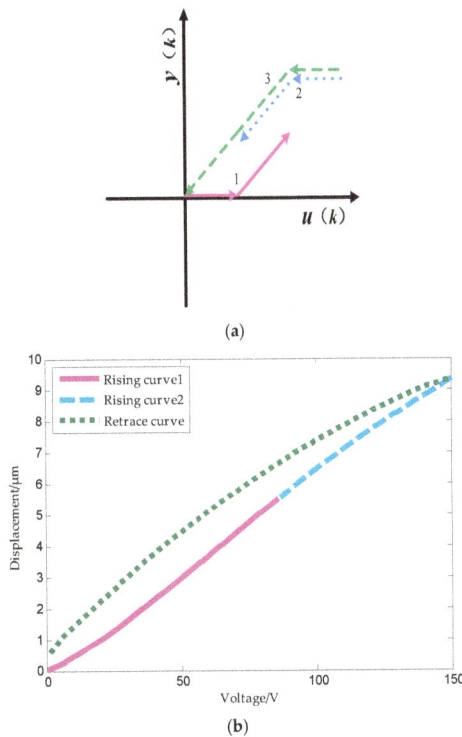

Figure 14. Relation of the play operator and the hysteresis curve.

The weight identification algorithm and threshold selection principle are consistent with Section 4.2. Table 2 shows the identification parameters of the tripartite PI model.

The tripartite PI model also includes the tripartite PI inverse model. Figure 15a,b show the modeling result of the tripartite PI model and its inverse model, respectively. Through MATLAB software modeling, we obtained the mean absolute error of the tripartite PI model as

$$\delta_3 = \frac{1}{n} \sum_{k=1}^{n} |e_k| = 0.02137 \ \mu\text{m}.$$

Table 2. Parameters of tripartite PI model. (*i* is the number of sampling point, r_1 is the threshold of the first stage play operator, w_1 is the weight of the first stage hysteresis operator in the mathematical sense; r_2 is the threshold of the second stage play operator, w_2 is the weight of the second stage hysteresis operator in the mathematical sense; r_3 is the threshold of the first stage play operator, w_3 is the weight of the first stage hysteresis operator in the mathematical sense).

i	r_1	w_1	r_2	w_2	r_3	w_3
1	0	0.0415	0	0.0529	0	0.0322
2	6.42	0.0097	15	0.0027	15	0.0081
3	12.84	0.0082	30	0.0067	30	0.0054
4	19.26	0.0067	45	0.0022	45	0.0044
5	25.68	0.0051	60	0	60	0.0065
6	32.10	0.0043	75	0	75	0.0039
7	38.52	0.0026	90	0	90	0.0087
8	44.94	0.0016	105	0	105	0.0016
9	51.36	0	120	0	120	0.0009
10	57.78	0	135	0	135	0.0008

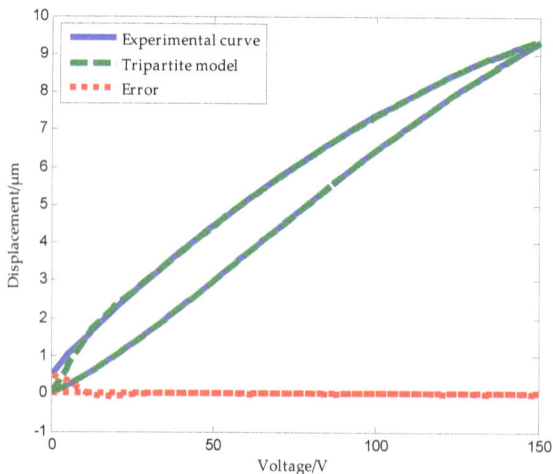

(**a**)

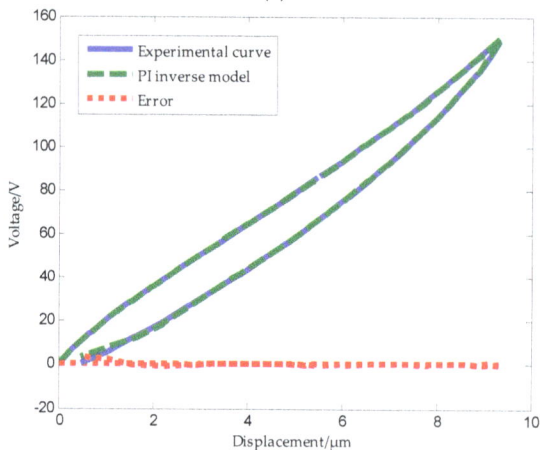

(**b**)

Figure 15. The tripartite PI model and its inverse model: (**a**) experimental and tripartite PI model hysteresis curves of a piezoelectric ceramic actuator; (**b**) experimental and tripartite PI inverse model hysteresis curves of a piezoelectric ceramic actuator.

It should be noted that the unilateral play operator used for modeling returns to the origin when the voltage comes back to zero, and the displacement of the tripartite PI model obtained by the weighted addition of the unilateral play operator is also forced to zero when the voltage drops to zero. Since the actual hysteresis curve does not return to zero when the voltage drops to zero, the tripartite PI model shows a high error when the voltage is close to 0 only during the process of reducing pressure.

5. Experiment Results and Discussion

The purpose of the microdisplacement positioning system is to make the expected piezoelectric ceramic output displacement equal to its actual displacement, as shown in Figure 16.

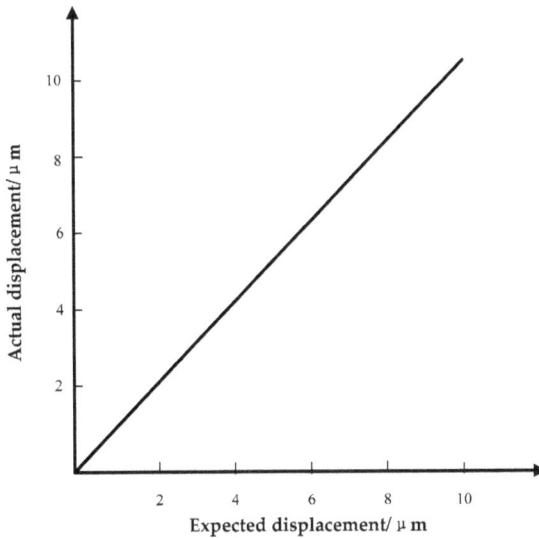

Figure 16. Desired relation of expected displacement and actual displacement.

Hence, in order to verify the effect of tripartite PI inverse model on the piezoelectric ceramic driver's hysteresis compensation, we designed a piezoelectric ceramic hysteresis model control effect comparison experiment. The PCA used for this study was the PSt150/4/7VS9 piezoelectric ceramic made by Core Tomorrow company (Harbin, China). We used the driving power HVA-150D. A3, made by Harbin Core Tomorrow Science & Technology Co., Ltd., to generate driving power (the input voltage was in the range of 0 to 150 V and the output displacement was in the range of 0 μm to 9.5 μm) to drive the piezoelectric ceramic and a Renishaw XL-80 laser interferometer to measure the displacement. The equipment used for the experiment was as shown in Figure 17. The driving power communicated with the host computer through the standard parallel port (SPP) parallel communication port. Here, we refer to the Preisach model parameters used in Song et al.'s study [16] and we used the experimental data in this study to establish the PCAs Preisach model. The desired displacement was taken as the input of the PI inverse model, the Preisach inverse model, and the tripartite PI inverse model. Thus, we got three sets of control voltages. These three sets of voltages were used to control the piezoelectric ceramic via the driving power. The output displacement was collected and recorded in real time by the laser interferometer. The control block diagram is shown in Figure 18. After the experiment, the data was processed and compared.

The results obtained are shown in Figure 19. The mean absolute errors (*MAE*) of the traditional inverse PI model, the Preisach inverse model, and the tripartite PI inverse model compensation controllers are $MAE = 0.19019$ μm, $MAE = 0.10893$ μm, and $MAE = 0.03549$ μm, respectively.

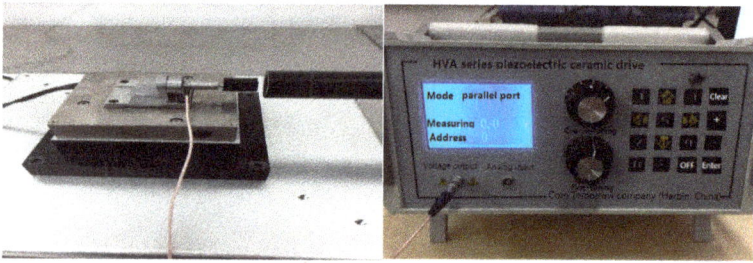

Figure 17. Experimental setup.

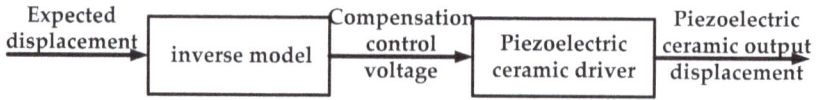

Figure 18. Block diagram of PI inverse model.

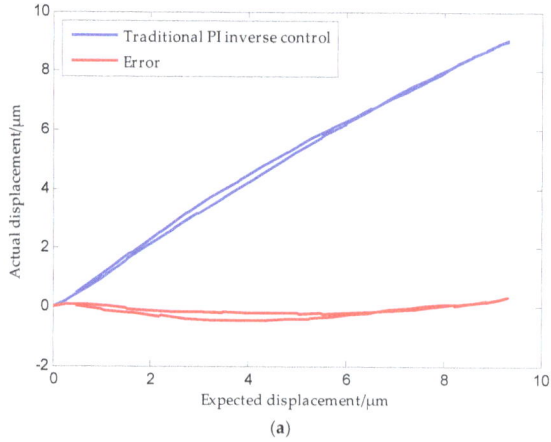

(a)

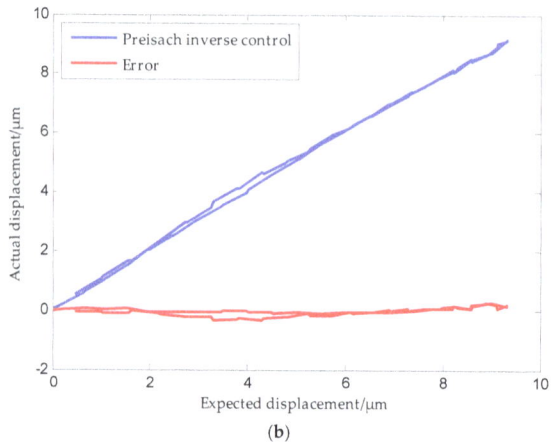

(b)

Figure 19. *Cont.*

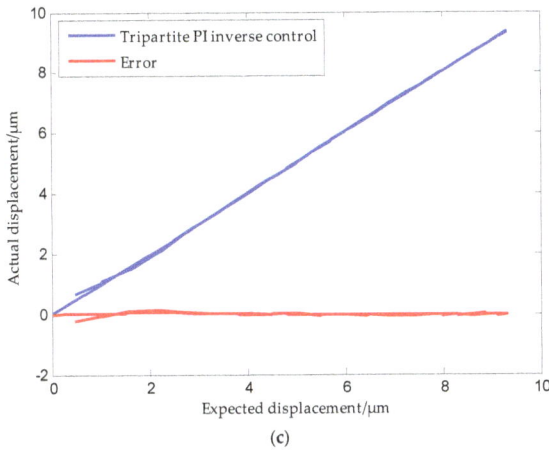

(c)

Figure 19. Positioning accuracies of three kinds of inverse models: (**a**) PI inverse model; (**b**) Preisach inverse model; (**c**) Tripartite PI inverse model.

Figure 20 shows a comparison of the positioning accuracies of the three models. It can be seen from the figure that the positioning accuracy of the tripartite PI model was higher than that of the traditional PI model and the Preisach model. Error analysis shows that the positioning accuracy was improved by more than 80% in the case of the tripartite inverse model when compared to the other two models. Experiments confirm that the proposed modeling method was effective.

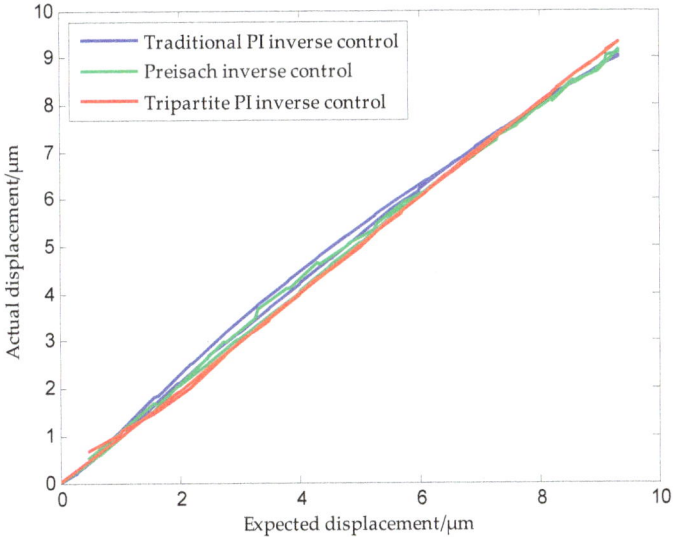

Figure 20. Comparison of the positioning accuracies of different models.

6. Conclusions

In this study, through the observation of the deformation rate for piezoelectric ceramics during the process of applying voltage, we arrived at the general law of the deformation rate of piezoelectric ceramics and this law has a certain universality. The tripartite PI model is proposed on the basis of the deformation rate law. The hysteresis curves of piezoelectric ceramics with different deformation rate laws are modeled to obtain the tripartite PI inverse model. Since the second segment is inverted, in the actual control application, the control voltage of the second stage needs to be used in reverse order. The tripartite PI modeling method does not introduce other parameters and other operators. The model is simple and easy to construct, and can accurately describe the characteristics of the main hysteresis loop. The tripartite PI inverse model has even more accurate precision as a series controller in the micro process and is used in the reverse cycle. Next, we will conduct a study about the remanent displacement of the hysteresis curve to broaden the scope of application of the model.

Acknowledgments: This study was supported by the Shenyang Jianzhu University (SJZU) Postdoctoral Innovation Fund Project (SJZUBSH201610); Discipline Content Education Project (XKHY2-26); National-Local Joint Engineering Laboratory of NC Machining Equipment and Technology of High-Grade Stone (SJSC-2015-13; SJSC-2016-3); Shenyang Jianzhu University Scientific Research Project (2017019); Colleges and Universities Basic Research Projects in Liaoning Province (2017).

Author Contributions: Dong An and Ying Xu designed and performed the experiments. Haodong Li analyzed the data. Lixiu Zhang wrote and reviewed the manuscript.

Conflicts of Interest: The authors declare no conflict of interest.

References

1. Tuma, T.; Lygeros, J.; Kartik, V.; Sebastian, A.; Pantazi, A. High-speed multiresolution scanning probe microscopy based on Lissajous scan trajectories. *Nanotechnology* **2012**, *23*, 185501. [CrossRef] [PubMed]
2. Braunsmann, C.; Schäffer, T.E. High-speed atomic force microscopy for large scan sizes using small cantilevers. *Nanotechnology* **2010**, *21*, 225705. [CrossRef] [PubMed]
3. Bhagat, U.; Shirinzadeh, B.; Tian, Y.; Zhang, D. Experimental analysis of laser interferometry-based robust motion tracking control of a flexure-based mechanism. *IEEE Trans. Autom. Sci. Eng.* **2013**, *10*, 267–275. [CrossRef]
4. Hansma, P.K.; Schitter, G.; Fantner, G.E.; Prater, C. High-speed atomic force microscopy. *Appl. Phys.* **2006**, *314*, 601–602.
5. Tuma, T.; Sebastian, A.; Lygeros, J.; Pantazi, A. The four pillars of nanopositioning for scanning probe microscopy: The position sensor, the scanning device, the feedback controller, and the reference trajectory. *Control Syst. IEEE* **2013**, *33*, 68–85. [CrossRef]
6. Clayton, G.M.; Tien, S.; Leang, K.K.; Zou, Q.; Devasia, S. A review of feedforward control approaches in nanopositioning for high-speed SPM. *J. Dyn. Syst. Meas. Control* **2009**, *131*, 636–650. [CrossRef]
7. Park, G.; Bement, M.T.; Hartman, D.A.; Smith, R.E.; Farrar, C.R. The use of active materials for machining processes: A review. *Int. J. Mach. Tools. Manuf.* **2007**, *47*, 2189–2206. [CrossRef]
8. Gozen, B.A.; Ozdoganlar, O.B. Design and evaluation of a mechanical nanomanufacturing system for nanomilling. *Precis. Eng.* **2012**, *36*, 19–30. [CrossRef]
9. Huang, S.; Tan, K.K.; Tong, H.L. Adaptive sliding-mode control of piezoelectric actuators. *IEEE Trans. Ind. Electron.* **2009**, *56*, 3514–3522. [CrossRef]
10. Zhu, W.; Rui, X. Hysteresis modeling and displacement control of piezoelectric actuators with the frequency-dependent behavior using a generalized Bouc–Wen model. *Precis. Eng.* **2016**, *43*, 299–307. [CrossRef]
11. Janaideh, A.; Farhan, M. Generalized Prandtl-Ishlinskii hysteresis model and its analytical inverse for compensation of hysteresis in smart actuators. *Mech. Ind. Eng.* **2009**, *9*, 307–312.
12. Li, P.; Li, P.; Sui, Y. Adaptive fuzzy hysteresis internal model tracking control of piezoelectric actuators with nanoscale application. *IEEE Trans. Fuzzy Syst.* **2016**, *24*, 1246–1254. [CrossRef]
13. Gu, G.Y.; Li, C.X.; Zhu, L.M.; Su, C.Y. Modeling and identification of piezoelectric-actuated stages cascading hysteresis nonlinearity with linear dynamics. *IEEE/ASME Trans. Mechatron.* **2016**, *21*, 1792–1797. [CrossRef]
14. Mokaberi, B.; Requicha, A.A.G. Compensation of Scanner Creep and Hysteresis for AFM Nanomanipulation. *IEEE Trans. Autom. Sci. Eng.* **2008**, *5*, 197–206. [CrossRef]
15. Gu, G.Y.; Zhu, L.M.; Su, C.Y. Integral resonant damping for high-bandwidth control of piezoceramic stack actuators with asymmetric hysteresis nonlinearity. *Mechatronics* **2014**, *24*, 367–375. [CrossRef]

16. Song, X.; Duggen, L.; Lassen, B.; Mangeot, C. Modeling and identification of hysteresis with modified preisach model in piezoelectric actuator. In Proceedings of the IEEE International Conference on Advanced Intelligent Mechatronics, Munich, Germany, 3–7 July 2017; pp. 1538–1543.

17. Cao, Y.; Chen, X.B. A survey of modeling and control issues for piezo-electric actuators. *J. Dyn. Syst. Meas. Control* **2015**, *137*, 14001. [CrossRef]

18. Jiles, D.C.; Atherton, D.L. Theory of ferromagnetic hysteresis (invited). *J. Magn. Mag.Mater.* **1986**, *61*, 48–60. [CrossRef]

19. Carrera, Y.; Avila-de La Rosa, G.; Vernon-Carter, E.J.; Alvarez-Ramirez, J. A fractional-order Maxwell model for non-Newtonian fluids. *Phys. A Stat. Mech. Its Appl.* **2017**, *482*, 276–285. [CrossRef]

20. Malczyk, R.; Izydorczyk, J. The frequency-dependent Jiles–Atherton hysteresis model. *Phys. B Condens. Matter* **2015**, *463*, 68–75. [CrossRef]

21. Liu, Y.; Liu, H.; Wu, H.; Zou, D. Modelling and compensation of hysteresis in piezoelectric actuators based on Maxwell approach. *Electron. Lett.* **2015**, *52*, 188–190. [CrossRef]

22. Liu, L.; Tan, K.K.; Chen, S.L.; Huang, S.; Lee, T.H. SVD-based Preisach hysteresis identification and composite control of piezo actuators. *ISA Trans.* **2012**, *51*, 430–438. [CrossRef] [PubMed]

23. Hassani, V.; Tjahjowidodo, T.; Do, T.N. A survey on hysteresis modeling, identification and control. *Mech. Syst. Signal Process.* **2014**, *49*, 209–233. [CrossRef]

24. Chen, H.; Tan, Y.; Zhou, X.; Dong, R.; Zhang, Y. Identification of dynamic hysteresis based on duhem model. In Proceedings of the International Conference on Intelligent Computation Technology and Automation, Shenzhen, China, 28–29 March 2011; pp. 810–814.

25. Lin, C.J.; Lin, P.T. Tracking control of a biaxial piezo-actuated positioning stage using generalized Duhem model. *Comput. Math. Appl.* **2012**, *64*, 766–787. [CrossRef]

26. Wang, G.; Chen, G.; Bai, F. Modeling and identification of asymmetric Bouc–Wen hysteresis for piezoelectric actuator via a novel differential evolution algorithm. *Sens. Actuators A Phys.* **2015**, *235*, 105–118. [CrossRef]

27. Huang, X.; Zeng, J.; Ruan, X.; Zheng, L.; Li, G. Structure, electrical and thermal expansion properties of PZnTe-PZT ternary system piezoelectric ceramics. *J. Am. Ceram. Soc.* **2017**, *101*, 274–282. [CrossRef]

28. Zhong, W. *Physics of Ferroelectrics*; Science Press: Beijing, China, 1996; pp. 294–297, 391–394.

29. Bridger, K.; Jones, L.; Poppe, F.; Brown, S.A.; Winzer, S.R. High-force cofired multilayer actuators. *Proc. SPIE* **1996**, *2721*, 341–352.

30. Lancée, C.T.; Souquet, J.; Ohigashi, H.; Bom, N. Transducers in medical ultrasound: Part One. Ferro-electric ceramics versus polymer piezoelectric materials. *Ultrasonics* **1985**, *23*, 138. [CrossRef]

31. Rabe, K.M.; Ahn, C.H.; Triscone, J.M. *Physics of Ferroelectrics*; Springer: Berlin/Heidelberg, Germany, 2007; pp. 203–234.

32. Merz, W.J. Domain Formation and Domain Wall Motions in Ferroelectric $BaTiO_3$ Single Crystals. *Phys. Rev.* **1954**, *95*, 690–698. [CrossRef]

33. Merz, W.J. Switching Time in Ferroelectric $BaTiO_3$ and Its Dependence on Crystal Thickness. *J. Appl. Phys.* **1956**, *27*, 938–943. [CrossRef]

micromachines

MDPI

Article

Modeling and Identification of the Rate-Dependent Hysteresis of Piezoelectric Actuator Using a Modified Prandtl-Ishlinskii Model

Yanding Qin [1,2,*], Xin Zhao [1,2] and Lu Zhou [1,2]

[1]   Institute of Robotics and Automatic Information System, Nankai University, Tianjin 300350, China; zhaoxin@nankai.edu.cn (X.Z.); zhoulu@nankai.edu.cn (L.Z.)
[2]   Tianjin Key Laboratory of Intelligent Robotics, Nankai University, Tianjin 300350, China
*   Correspondence: qinyd@nankai.edu.cn; Tel.: +86-22-2350-5960

Academic Editors: Ulrich Schmid and Michael Schneider
Received: 20 December 2016; Accepted: 29 March 2017; Published: 4 April 2017

Abstract: Piezoelectric actuator (PEA) is an ideal microscale and nanoscale actuator because of its ultra-precision positioning resolution. However, the inherent hysteretic nonlinearity significantly degrades the PEA's accuracy. The measured hysteresis of PEA exhibits strong rate-dependence and saturation phenomena, increasing the difficulty in the hysteresis modeling and identification. In this paper, a modified Prandtl-Ishlinskii (PI) hysteresis model is proposed. The weights of the backlash operators are updated according to the input rates so as to account for the rate-dependence property. Subsequently, the saturation property is realized by cascading a polynomial operator with only odd powers. In order to improve the efficiency of the parameter identification, a special control input consisting of a superimposition of multiple sinusoidal signals is utilized. Because the input rate of such a control input covers a wide range, all the parameters of the hysteresis model can be identified through only one set of experimental data, and no additional curve-fitting is required. The effectiveness of the hysteresis modeling and identification methodology is verified on a PEA-driven flexure mechanism. Experimental results show that the modeling accuracy is on the same order of the noise level of the overall system.

Keywords: piezoelectric actuator; hysteresis; rate-dependence; Prandtl-Ishlinskii

1. Introduction

Piezoelectric actuator (PEA) has been widely utilized in ultra-precision positioning and manipulation applications due to its sub-nano motion resolution, high output force and fast response capabilities [1]. However, the disadvantages of the PEA are also distinct: (1) the PEA can be easily damaged by large bending torques or external impacts as the material is brittle; (2) the stroke of the PEA is very limited. The ratio between the stroke and the length of the PEA is typically on the level of 10 μm/cm; and (3) the inherent rate-dependent hysteretic and creeping nonlinearities significantly degrade the PEA's motion accuracy. In practice, flexure-based displacement amplification mechanisms are generally adopted to magnify the stroke of the PEA, such as the flexural lever mechanism and the flexural Scott-Russell mechanism [2,3]. Capacity-based, laser-based sensors and strain gauges are generally utilized to measure such small displacements. For the motion control of PEAs, the hysteresis can be compensated using either the modeling-inversion based approaches [4–6] or the model-free feedback control [7]. Unlike the hysteresis, the creep is the slow drift of the PEA's output over time that can be easily compensated through the feedback control.

Hysteresis modeling and compensation have been extensively investigated in recent decades. One widely employed hysteresis model is the Preisach model [8–10] which describes the hysteresis phenomenon through a double integral. The Preisach model can be used for hysteresis compensation and very high motion accuracy can be achieved if it is combined with other feedback controllers [11]. As the modeling accuracy of the Preisach model is highly related to the segmentation of the α-β plane, one needs to increase the model's order to obtain higher modeling accuracy. Another widely employed hysteresis model is the Prandtl-Ishlinskii (PI) model [12,13]. The PI model is becoming more and more popular due to its simplicity in formulation, high modeling accuracy, and theoretical reversibility in the rate-independent form, making it attractive in real-time implementations. It must be noted that the classical PI model is static and symmetrical about the loop center. However, the measured hysteresis of the PEA exhibits strong rate-dependence and asymmetry (saturation) properties. Rate-dependence is the phenomenon in which the measured hysteresis curve of a PEA will become wider with the increment of the input rate. And the measured hysteresis curve of a PEA is not strictly symmetrical about its loop center, which is defined as the saturation property. These factors significantly increase the difficulty in hysteresis modeling and compensation. In literature, different modifications have been made to the classical PI model to better fit the saturation and rate-dependence properties of the measured hysteresis of PEAs [12,14–16].

The strong couplings between the hysteretic and creeping nonlinearities and the linear dynamics of a PEA make it impossible to isolate the hysteretic nonlinearity from its linear dynamics. Further, the output of a PEA is also susceptible to many factors, such as the preload force, the external load, and the dynamics of the transmission chains. This makes it difficult to precisely predict the behavior of a PEA if only a hysteresis model or a dynamics model is constructed. Therefore, an integrated model of both the linear dynamics and the nonlinear hysteresis will significantly improve the modeling accuracy of a PEA. In the research work of Hassani and Tjahjowidodo [17], both the dynamics of the mechanism and the hysteresis of the PEA are modeled and integrated together as a full hysteresis-dynamics model, where the hysteretic response of the PEA is adopted as the input to the linear dynamics of the mechanism. The combination of both the linear dynamics and nonlinear hysteresis obviously increases the modeling accuracy of the overall system.

The PEA's dynamics is very important in the scanning- or vibration-based applications where the PEA moves very fast, such as the atomic force microscope and ultrasonic motor. However, in many micro and nano scale manipulations, such as in the manipulation and characterization of living cells [18,19], the endeffector follows the motion of the master operator's hand, or moves very slowly, typically on the order of several Hertz. For these very slow motions, the PEA's dynamics is not obvious and the PEA's hysteresis becomes the dominant factor affecting the behavior of the overall system. Therefore, the hysteresis modeling and compensation is important to improve the performance of such systems. This paper focuses on the hysteresis modeling and identification of such systems. In our previous work [20], the saturation property was accounted for by the use of a polynomial operator, and the rate-dependence property was accounted for by varying the weight vector of the PI model according to the input rate. Although very high modeling accuracy was achieved, the threshold vector was still manually assigned. As a result, a trial and error process was inevitable, and a high level of knowledge on the characteristics of the PEA's hysteresis was required. From the practitioner's point of view, it is desirable to eliminate such a complex modeling and identification process to achieve ease of use in real implementations. In order to improve the applicability of the hysteresis modeling and compensation method proposed in our previous work [20], this paper aims to eliminate all the manual interventions during the parameter identification process. As a result, one only has to check the bounds of the input range from the manual of the PEA and select the order of the hysteresis model, and no other post processing or manual intervention is required during the parameter identification process.

2. Materials and Methods

2.1. Materials

A three degrees-of-freedom (DOF) flexure mechanism presented in [21] is utilized for the hysteresis modeling and verification in this paper. This mechanism is actuated by three PEAs (Model

PZS001 from Thorlabs (Newton, NJ, USA)). The input range of the PEA is 0–10 V and the maximum displacement of the PEA is 11.6 ± 2.0 μm. The displacement of the PEA is measured by the strain gauges attached on the PEA in a full Wheatstone bridge configuration. The control voltages are exerted on the PEA through a piezo driver (Model MDT693B from Thorlabs). The data acquisition task is implemented on a PXI platform (Model 1082 equipped with a PXI-8135 controller, a PXIe-6363 data acquisition card and a TB-4330 bridge amplifier, all from National Instruments (Austin, TE, USA)) and runs in the real-time environment of Labview (Version 2014 SP1, National Instruments). As shown in Figure 1, the overall system is mounted on an optical table to isolate the ground disturbances. The noise level of the system is measured to be 100 nm. For the parameter identification and validation in this paper, different control signals are exerted on the PEA in one axis of the mechanism and the resultant extension of the PEA is measured by the strain gauges.

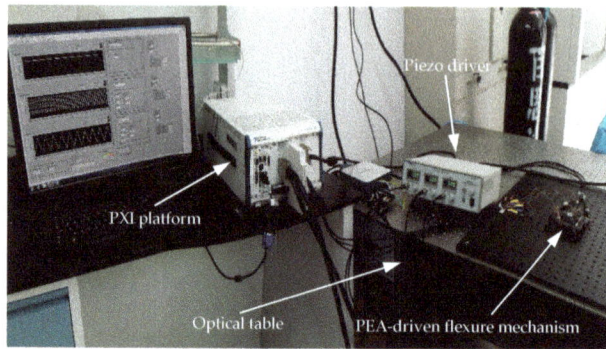

Figure 1. The experimental setup of the overall system.

2.2. Modified Prandtl-Ishlinskii Hysteresis Model

2.2.1. Classical Prandtl-Ishlinskii Model

The basic component of all PI-based hysteresis models is the backlash operator in the following formulation:

$$H_r(u,t) = \max\{x(t) - r, \min\{u(t) + r, H_r(t-T)\}\}$$
$$H_r(u,0) = \max\{u(0) - r, \min\{u(0) + r, 0\}\}$$

(1)

where $H_r(u,t)$ denotes the backlash operator, $u(t)$ and $y(t)$ represent the input and output of the backlash operator, respectively, r is the threshold of the backlash operator, t is the current time, and the system runs with a sampling period of T. The initial condition can be set to zero as a PEA is typically activated from its de-energized state.

The classical PI model is defined as the weighted superposition of n backlash operators, i.e.,

$$z(t) = \sum_{i=1}^{n} w_i H_{ri}(u,t) = [w_1, w_2, ..., w_n] \cdot [H_{r1}(u,t), H_{r2}(u,t), ..., H_{rn}(u,t)]^T = \mathbf{w}^T \cdot \mathbf{H_r}(u,t)$$

(2)

where $z(t)$ is the output of the classical PI model, n is defined as the order of the PI model, $\mathbf{w} = [w_1, w_2, \ldots, w_n]^T$ and $\mathbf{H_r}(u,t) = [H_{r1}(u,t), H_{r2}(u,t), \ldots, H_{rn}(u,t)]^T$ are the weight vector and the backlash operator vector, respectively.

2.2.2. Modeling of the Saturation Property

It is noted that the classical PI model shown in Equation (2) is rate-independent and symmetrical about its loop center. However, the measured hysteresis loops of PEAs are rate-dependent and asymmetric (saturation property). In order to improve the modeling accuracy, a modified PI model is proposed and schematically illustrated in Figure 2a, where an additional saturation operator is cascaded to the classical PI model. Following the notations in Equation (2), $H_{rn}(\bullet)$ stands for the backlash operator in the classical PI model and w_n is the weight vector in the backlash operator. A special polynomial operator with only odd powers in the following formulation is utilized as the saturation operator:

$$\hat{y}(t) = S[z](t) = c_1 z(t) + c_3 z^3(t) + \cdots + c_m z^m(t), \quad m = 1,3,5,\cdots \tag{3}$$

where $S[z](t)$ denotes the saturation operator, $\hat{y}(t)$ is the output, c_i (i = 1, 3, 5, ..., m) is the coefficients of the polynomial, and m is the order of the polynomial.

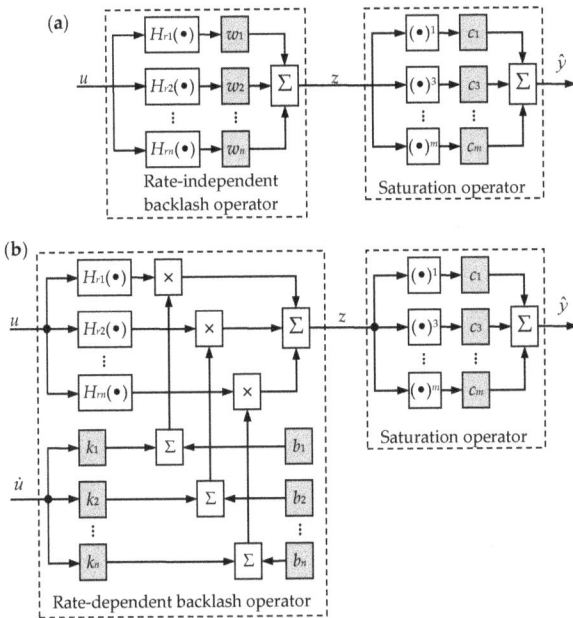

Figure 2. Schematic diagram of the modified Prandtl-Ishlinskii (PI) model: (**a**) rate-independent hysteresis model and (**b**) rate-dependent hysteresis model.

Unlike the common polynomials, all the even powers and the constant are totally eliminated to guarantee the axial symmetry of the saturation operator about the origin. Through literature review, one-sided dead-zone operator is another popular saturation operator [14,22]. However, this operator is piecewise in nature, even if the operator's order increases. On the contrary, the coefficients of the polynomial are fewer but the curve of the polynomial is smooth, as verified in our previous work [20]. This helps to increase the modeling accuracy while decreasing the model complexity. In addition, more complex saturation property is possible by tuning the degree of the polynomial.

Substituting Equation (2) into Equation (3), the rate-independent hysteresis model is written as:

$$\hat{y}(t) = S\left[\mathbf{w}^T \cdot \mathbf{H_r}(u)\right](t) \tag{4}$$

2.2.3. Modeling of the Rate-Dependence Property

The shape of the PEA's hysteresis loop varies with different control inputs, e.g., the hysteresis loop become thicker if the rate or the frequency of the input increases. For the modeling of the rate-dependence, it has been verified that linearly tuning the weight vector of the modified PI model according to the input rate can significantly improve the modeling accuracy. The schematic diagram of the rate-dependent hysteresis model is given in Figure 2b. The linear relationship can be expressed in the following equation:

$$\mathbf{w} = \mathbf{k} \cdot \dot{u}(t) + \mathbf{b} \tag{5}$$

where $\mathbf{k} = [k_1, k_2, ..., k_n]^T$, $\mathbf{b} = [b_1, b_2, ..., b_n]^T$, k_i and b_i (i = 1, 2, ..., n) are the slop and offset vectors, respectively, and $\dot{u}(t)$ is the input rate.

Substituting Equations (2) and (5) into Equation (3), the rate-dependent hysteresis model is obtained:

$$Y(t) = S\left\{ \left[\mathbf{k} \cdot \dot{u} + \mathbf{b}\right]^T \cdot \mathbf{H_r}(u) \right\}(t) \tag{6}$$

A similar modeling approach using a third order polynomial as the saturation operator was also proposed in [23], where the classical PI model and the saturation operator are connected in parallel. It is also noted that the hysteresis model developed in [23] is rate-independent. In our approach, the saturation operator is cascaded to the classical PI model in series, and the order of the polynomial can be tuned for more complex hysteresis. More importantly, the hysteresis model defined in Equation (6) is rate-dependent, and thus the modeling accuracy can be guaranteed.

2.3. Full Parameter Identification

The parameters that need to be identified include the threshold and weight vectors, and the coefficients of the saturation operator. The threshold vector is very important in the identification. Higher modeling accuracy can be achieved if a higher order of backlash operators and fine spacing are selected. However, higher order will increase the model complexity, and the computation time will also increase significantly, causing severe problems in real-time applications. A trade-off has to be made between the modeling accuracy and the system complexity. One practical solution is assigning fine spacing at low threshold values while assigning coarse spacing at larger threshold values. As a result, a threshold vector with 10th order or above is adequate to achieve satisfactory results. However, through literature review, this non-uniform spacing is typically assigned manually through a laborious trial and error process, and thus the prior experience on hysteresis modeling is highly demanded. This significantly affects the applicability of the PI-based approaches.

In our previous work [20], the threshold vector is manually assigned, and the saturation and backlash operators are identified separately for a shorter computation time. Thus, two sets of experimental data are required in the identification. In this paper, a highly efficient full parameter identification approach is proposed to identify all the parameters through only one set of experimental data. The practitioner only has to find out the input range of the system from the manual. No other prior experience on the PEA's hysteresis or post processing, such as curve-fitting, is required. This guarantees ease of use and high efficiency, and thus signifies progress from our previous approach [20].

2.3.1. Error Functions for Parameter Identification

When the PEA moves very slowly, e.g., tracking a trajectory below 1 Hz or following the trajectory of a master operator, the rate-dependence is negligible, resulting in a rate-independence (static) hysteresis. In this case, the rate-independent model defined in Equation (4) is sufficient to predict the output of the PEA. All the parameters in Equation (4) can be identified by comparing the model output with the measured hysteresis and minimizing the following error function:

$$E[\hat{y}, y](\mathbf{r}, \mathbf{w}, \mathbf{c}, t) = \hat{y}(t) - y(t) = S\left[\mathbf{w}^T \cdot \mathbf{H_r}\right](t) - y(t) \tag{7}$$

where $\mathbf{r} = [r_1, r_2, ..., r_n]^T$ is the threshold vector.

If the PEA moves fast, the rate-dependence will become very obvious. In this case, the rate-dependent model defined in Equation (6) should be used to predict the output of the PEA.

Similarly, all the parameters in Equation (6) can be identified by comparing the model output with the measured hysteresis and minimizing the other error function:

$$E[\hat{y},y](\mathbf{r},\mathbf{k},\mathbf{b},\mathbf{c},t) = \hat{y}(t) - y(t) = S\Big\{[\mathbf{k}\cdot\dot{u} + \mathbf{b}]^T\cdot\mathbf{H_r}\Big\}(t) - y(t) \tag{8}$$

In the parameter identification process, the method of least squares is adopted to minimize the error functions defined in Equations (7) and (8). The parameter identification is implemented in the environment of MATLAB (Version R2014a, MathWorks, Natick, MA, USA) and the function *lsqcurvefit* is selected.

2.3.2. Input Signals for Parameter Identification

For the rate-dependent hysteresis identification, it is straightforward to excite the system using control signals at different constant input rates (e.g., saw tooth signals with different slopes) and to identify the parameters in each case separately, as proposed in the research work of Ang et al. [14]. Subsequently, an additional curve fitting is conducted in order to obtain a general model that covers a certain range of input rates. This process is laborious as one has to make many measurements so as to obtain adequate data for the identification. Further, as saw tooth signals are not consistent, the modal vibrations of the PEA-driven system are likely to be excited by the high frequency components of the saw tooth signal.

Alternately, it is possible to identify all the parameters through only one set of experimental data if the input rate of the control signal is not constant but spans a certain range. Based on our previous work, the superimposition of multiple sinusoidal signals at different frequencies in the following form is a better alternative:

$$u(t) = \sum_N A_i \sin\Big(2\pi f_i t - \frac{\pi}{2}\Big), \quad N \geq 2 \tag{9}$$

where N is the number of the sinusoidal signals, A_i and f_i are the magnitude and frequency of the sinusoidal signal, respectively.

The superimposition of multiple sinusoidal signals is superior to the saw tooth signal in that: (1) the input rate can span a wide range through a careful selection of the sinusoidal signals, and (2) the signal is consistent and the modal vibration can be avoided.

2.3.3. Non-Uniform Initialization of the Threshold Vector

As previously stated in this paper, non-uniform spacing of the threshold vector can achieve better modeling accuracy. As a result, during the parameter identification process, the threshold vector is initialized using a cubic relationship to guarantee fine spacing at small threshold values and coarse spacing at larger threshold values:

$$r_i = \Big(\frac{i}{n}\Big)^3\cdot\frac{U}{2}, \quad i = 1,2,...,n \tag{10}$$

where U is the upper bound of the input signal.

3. Results

3.1. Rate-Independent Hysteresis Identification

For the identification of the rate-independent hysteresis, a 10 Vp-p, 1 Hz sinusoidal signal is adopted as the control input to the system. The measured displacement of the PEA and the model output of the rate-independent hysteresis model in Equation (4) are plotted in Figure 3. It can be observed that the identified hysteresis model agrees with the measured hysteresis of the PEA. The modeling error is 5.287 $\pm$ 62.41 nm. The identified parameters are given below:

$$\mathbf{r} = \begin{bmatrix} 0,\ 4.933e^{-3},\ 0.02294,\ 0.1830,\ 0.4944,\ 1.109,\ 1.839,\ 2.617,\ 3.779,\ 4.922 \end{bmatrix}^T$$
$$\mathbf{w} = \begin{bmatrix} -2.31,\ 2.137,\ 0.35,\ 9.3e^{-3},\ 0.02532,\ 0.02235,\ 0.01895,\ 0.01826,\ 0.02765,\ -0.01929 \end{bmatrix}^T \quad (11)$$
$$\mathbf{c} = \begin{bmatrix} -8.794e^{-4},\ 0,\ -0.05066,\ 0,\ 4.818 \end{bmatrix}^T$$

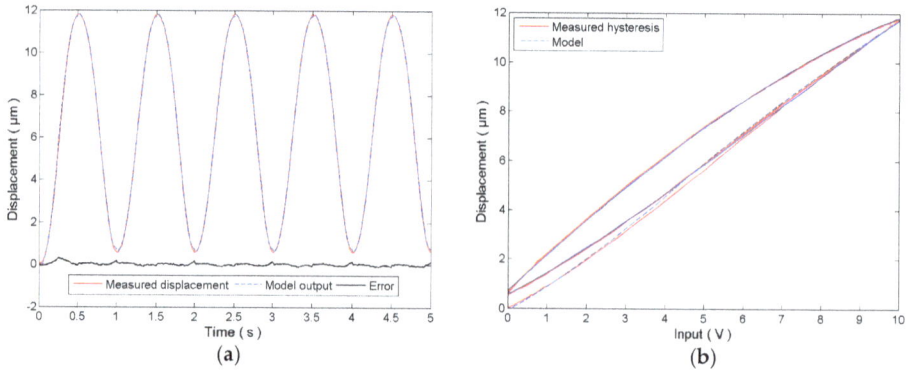

Figure 3. Identification of the rate-independent hysteresis model using a 1 Hz sinusoidal signal: (**a**) time plot and (**b**) hysteresis plot.

3.2. Rate-Dependent Hysteresis Identification

In the identification of the rate-dependent hysteresis, the superimposition of two sinusoidal signals is chosen where $A_1 = 2, f_1 = 10$, and $A_2 = 3, f_2 = 5$, respectively. This control signal is then exerted on the PEA. The parameters of the rate-dependent hysteresis model are identified according to the error function in Equation (8). The measured displacement of the PEA and the model output are plotted in Figure 4. Similar to the results in the rate-independence case, the identified model follows the measured hysteresis of the PEA well with a modeling error of 3.526 ± 45.55 nm. The identified parameters are given below:

$$\mathbf{r} = \begin{bmatrix} 0, 1.772e^{-3},\ 0.03242,\ 0.08513,\ 0.1940,\ 0.8016,\ 1.356,\ 2.104,\ 3.055,\ 3.438 \end{bmatrix}^T$$
$$\mathbf{k} = \begin{bmatrix} -0.1041, 0.1075,\ -3.254e^{-3},\ -7.082e^{-6},\ -1.974e^{-4},\ -9.755e^{-6},\ 1.232e^{-6},\ -2.045e^{-5},\ 4.844e^{-5},\ -5.606e^{-5} \end{bmatrix}^T$$
$$\mathbf{b} = \begin{bmatrix} -0.5510,\ 0.6097,\ -0.01653,\ 0.01098,\ 1.262e^{-3},\ 4.928e^{-3},\ 4.950e^{-3},\ 6.183e^{-3},\ 6.393e^{-3},\ -3.426e^{-4} \end{bmatrix}^T \quad (12)$$
$$\mathbf{c} = \begin{bmatrix} 0.5682,\ 0,\ -3.391,\ 0,\ 17.54 \end{bmatrix}^T$$

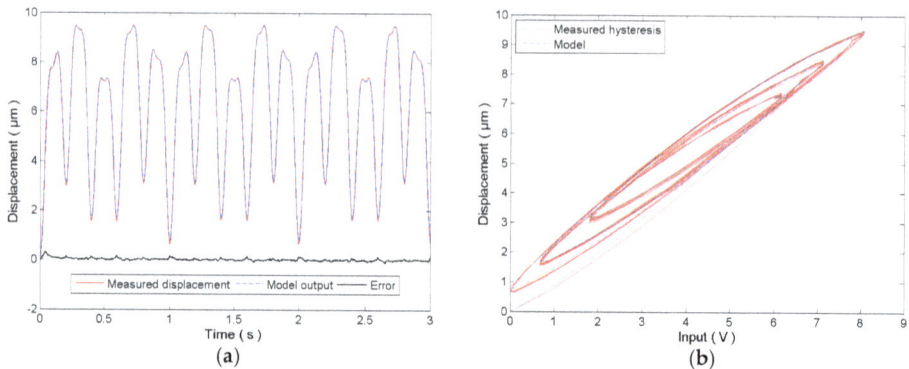

Figure 4. Identification of the rate-dependent hysteresis model using the superimposition of two sinusoidal signals: (**a**) time plot and (**b**) hysteresis plot.

3.3. Verifications of the Rate-Independent and Rate-Dependent Models

The identified rate-independent hysteresis model is verified using a 10 Vp-p, 1 Hz triangle signal. The experimental results are given in Figure 5. It is observed that the model output follows the measured displacement well and the estimation error is measured to be 79.56 ± 77.6 nm, on the same order of the noise level of the system. Therefore, the rate-independent is applicable for slow trajectories, e.g., below 1 Hz. Taking the relatively simple structure into consideration, the rate-independent hysteresis model is a better choice if the system moves slowly.

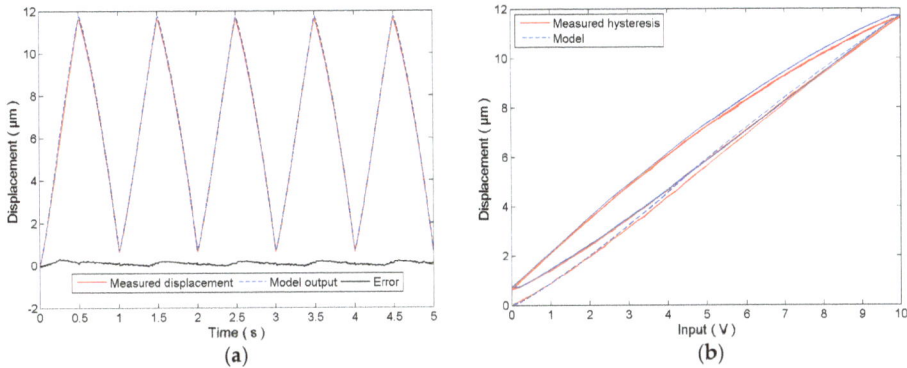

Figure 5. Verification of the rate-independent hysteresis model using another 1 Hz triangular signal: (**a**) time plot and (**b**) hysteresis plot.

For higher frequencies, both the identified rate-independent and rate-dependent hysteresis models are verified using an input signal consisting of four sinusoidal signals with $A_1 = 1, f_1 = 10, A_2 = 1, f_2 = 7.5, A_3 = 1, f_3 = 5, A_4 = 2$, and $f_4 = 4$. The measured displacement and the model outputs are given in Figure 6a. It can be observed that both the rate-independent and rate-dependent models can follow the measured displacements well. The estimation errors are measured to be 88.10 ± 86.29 nm and -48.77 ± 57.22 nm for the rate-independent and rate-dependent models, respectively. The estimation error of the rate-independent model is slightly higher than its modeling error. This is reasonable as the rate-independent model is identified using a 1 Hz sinusoidal signal and focuses mainly on low-frequency signals. The high-frequency components of the superimposed signal will definitely affect the accuracy of the rate-independent model. On the contrary, the estimation error of the rate-dependent model is on the same order of the modeling error. The error plot in Figure 6b clearly shows that the rate-dependent model achieves better performance than the rate-independent model for fast trajectories.

Because the superimposed sinusoidal signal in Figure 6 only consists of four frequencies, experiments are further conducted to test the performance of the identified rate-dependent model using a 0.1–20 Hz swept sinusoidal signal. Compared with the superimposed sinusoidal signal, the swept sinusoidal signal is also smooth but it contains all the frequency components between 0.1 Hz and 20 Hz, and thus the overall performance of the rate-dependent model over a wider frequency range can be examined. The experimental results are given in Figure 7. It is observed that the rate-dependent model can still follow the measured displacement well, as observed in the zoomed-in insets in Figure 7. However, since the maximum frequency component in the identification is only 10 Hz, the modeling accuracy for higher frequencies is not guaranteed. Experimental results in Figure 7 show that the estimation error will become larger for higher frequencies. The maximum estimation error in the last one second is 94.11 nm, corresponding to 8.25% of the measured displacement. Therefore, the identified rate-dependent model is applicable for fast trajectories with a frequency range of 20 Hz.

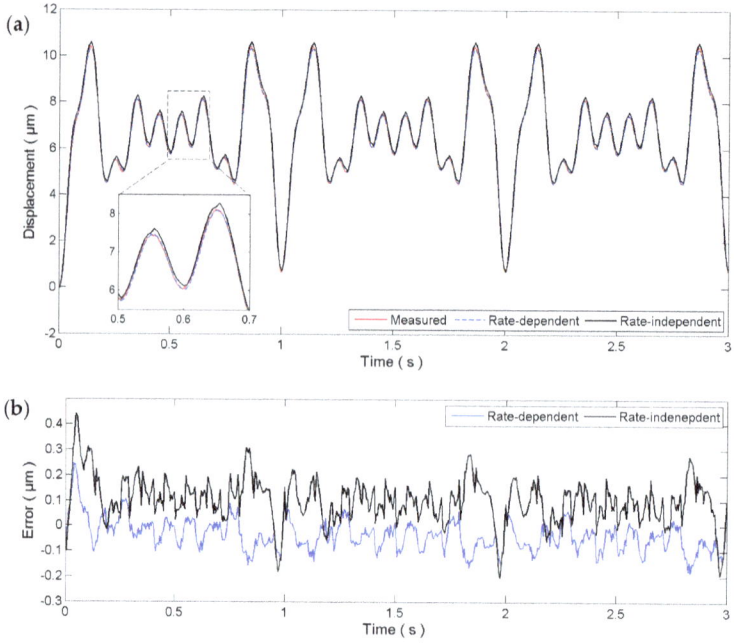

Figure 6. Verification of the identified hysteresis models using a superimposition of four sinusoidal signals: (**a**) time plot and (**b**) error plot.

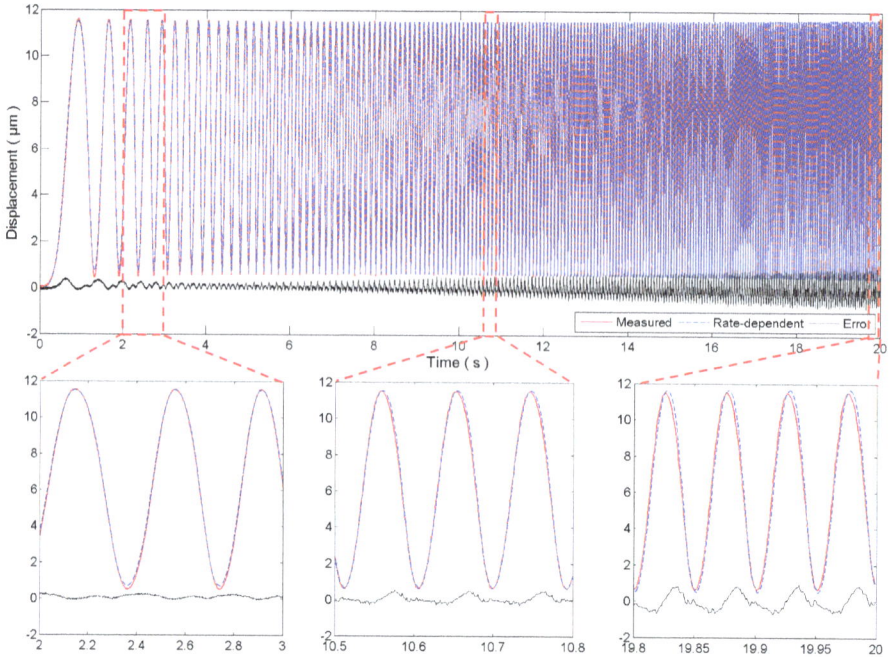

Figure 7. Verification of the rate-dependent model using a 0.1–20 Hz swept sinusoidal signal.

4. Discussion

The hysteresis modeling and compensation has become an important issue in the motion control of PEAs. The rate-dependence and saturation (asymmetry) phenomena are observed in the measured hysteresis curves of PEAs. In this paper, the PI model is selected to build the hysteresis model. Two important modifications are made to the classical PI model: (1) the weights of the backlash operators are dynamically updated according to the change in the input rate so as to account for the rate-dependence, and (2) a polynomial operator with only odd powers is cascaded to the backlash operators to adjust the shape of the hysteresis loop to model the saturation property.

The parameters that need to be identified include the threshold and weight vectors of the backlash operators and the coefficients of the polynomial operator. The efficiency of the conventional parameter identification approach is low because of the huge amount of experimental data and the time-consuming post processes, such as the curve fitting. Further, the sudden change in the input signal might excite higher order modal vibrations of the system during the experiment. Another problem in the parameter identification is the manual intervention. For instance, the threshold vector is generally manually assigned to guarantee fine spacing at small values and coarse spacing at larger values. This task requires rich experience on the hysteresis modeling of PEAs, and thus is not practical for beginners or common practitioners with limited experience. Therefore, it is necessary to simplify the parameter identification to achieve minimal manual intervention.

A full parameter identification approach is proposed in this paper. The basic idea is to choose a special input signal covering a wide range of input rates so that the response of the PEA to different input rates can be obtained in one single measurement. The superimposition of multiple sinusoidal signals with different frequencies is an ideal input signal as it covers a wide range of input rates. More importantly, the input signal is smooth and will not excite the higher modal vibration of the system. Subsequently, the method of least squares is utilized to identify all the parameters automatically without any manual intervention. This methodology is superior in that prior experience on hysteresis modeling is not required any more, thus guaranteeing ease of use even for the beginner.

The effectiveness of the proposed methodology is verified on a PEA-driven flexure mechanism. Both rate-independent and rate-dependent hysteresis identifications are conducted. Experimental results show that the rate-independent model is adequate to describe the motion of the PEA if the PEA works in slowly moving scenarios, such as following the trajectory of a master operator. For the rate-dependent hysteresis model, experimental results show that the modeling accuracy is high in the frequency range of 20 Hz.

Acknowledgments: This work is supported by the National Natural Science Foundation of China under Grants 61403214 and 61327802, in part by the Natural Science Foundation of Tianjin under Grants 14JCZDJC31800, 14ZCDZGX00801, and 14JCQNJC04700.

Author Contributions: Y.Q. and X.Z. conceived and designed the experiments; Y.Q. and L.Z. performed the experiments; Y.Q. analyzed the data and wrote the paper.

Conflicts of Interest: The authors declare no conflict of interest.

References

1. Clark, L.; Shirinzadeh, B.; Tian, Y.; Oetomo, D. Laser-based sensing, measurement, and misalignment control of coupled linear and angular motion for ultrahigh precision movement. *IEEE/ASME Trans. Mechatron.* **2015**, *20*, 84–92. [CrossRef]
2. Qin, Y.; Shirinzadeh, B.; Zhang, D.; Tian, Y. Design and kinematics modeling of a novel 3-DOF monolithic manipulator featuring improved Scott-Russell mechanisms. *ASME J. Mech. Des.* **2013**, *135*, 101004. [CrossRef]
3. Yong, Y.K.; Aphale, S.S.; Moheimani, S.O.R. Design, identification, and control of a flexure-based XY stage for fast nanoscale positioning. *IEEE Trans. Nanotechnol.* **2009**, *8*, 46–54. [CrossRef]
4. Leang, K.K.; Devasia, S. Design of hysteresis-compensating iterative learning control for piezo-positioners: Application to atomic force microscopes. *Mechatronics* **2006**, *16*, 141–158. [CrossRef]

5. Janaideh, M.A.; Su, C.-Y.; Rakheja, S. Development of the rate-dependent Prandtl-Ishlinskii model for smart actuators. *Smart Mater. Struct.* **2008**, *17*, 035026. [CrossRef]

6. Rakotondrabe, M. Bouc-wen modeling and inverse multiplicative structure to compensate hysteresis nonlinearity in piezoelectric actuators. *IEEE Trans. Autom. Sci. Eng.* **2011**, *8*, 428–431. [CrossRef]

7. Zhong, J.; Yao, B. Adaptive robust precision motion control of a piezoelectric positioning stage. *IEEE Trans. Control Syst. Technol.* **2008**, *16*, 1039–1046. [CrossRef]

8. Xiao, S.; Li, Y. Modeling rate-dependent and thermal-drift hysteresis through preisach model and neural network optimization approach. In *Advances in Neural Networks, Part 1, LNCS*; Springer: Berlin/Heidelberg, Germany, 2012; Volume 7367, pp. 179–187.

9. Wang, H.; Wilksch, J.J.; Strugnell, R.A.; Gee, M.L. Role of capsular polysaccharides in biofilm formation: An AFM nanomechanics study. *ACS Appl. Mater. Interfaces* **2015**, *7*, 13007–13013. [CrossRef] [PubMed]

10. Xiao, S.; Li, Y. Modeling and high dynamic compensating the rate-dependent hysteresis of piezoelectric actuators via a novel modified inverse preisach model. *IEEE Trans. Control Syst. Technol.* **2013**, *21*, 1549–1557. [CrossRef]

11. Tang, H.; Li, Y. Feedforward nonlinear PID control of a novel micromanipulator using preisach hysteresis compensator. *Robot. Comput. Integr. Manuf.* **2015**, *34*, 124–132. [CrossRef]

12. Chen, Y.; Qiu, J.; Palacios, J.; Smith, E.C. Tracking control of piezoelectric stack actuator using modified Prandtl-Ishlinskii model. *J. Intell. Mater. Syst. Struct.* **2013**, *24*, 753–760. [CrossRef]

13. Zhang, Y.; Yan, P. Sliding mode disturbance observer-based adaptive integral backstepping control of a piezoelectric nano-manipulator. *Smart Mater. Struct.* **2016**, *25*, 125011. [CrossRef]

14. Ang, W.T.; Khosla, P.K.; Riviere, C.N. Feedforward controller with inverse rate-dependent model for piezoelectric actuators in trajectory-tracking applications. *IEEE/ASME Trans. Mechatron.* **2007**, *12*, 134–142. [CrossRef]

15. Bashash, S.; Jalili, N. Robust multiple frequency trajectory tracking control of piezoelectrically driven micro/nanopositioning systems. *IEEE Trans. Control Syst. Technol.* **2007**, *15*, 867–878. [CrossRef]

16. Janaideh, M.A.; Rakheja, S.; Su, C.-Y. A generalized Prandtl-Ishlinskii model for characterizing the hysteresis and saturation nonlinearities of smart actuators. *Smart Mater. Struct.* **2009**, *18*, 045001. [CrossRef]

17. Hassani, V.; Tjahjowidodo, T. Structural response investigation of a triangular-based piezoelectric drive mechanism to hysteresis effect of the piezoelectric actuator. *Mech. Syst. Signal Process.* **2013**, *36*, 210–223. [CrossRef]

18. Li, M.; Dang, D.; Liu, L.; Xi, N.; Wang, Y. Imaging and force recognition of single molecular behaviors using atomic force microscopy. *Sensors* **2017**, *17*, 200. [CrossRef] [PubMed]

19. Yang, R.; Liu, L.; Zhang, C.; Xi, N.; Yang, J. Investigation of penetration using atomic force microscope: Potential biomarkers of cell membrane. *Micro Nano Lett.* **2015**, *10*, 248–252. [CrossRef]

20. Qin, Y.; Tian, Y.; Zhang, D.; Shirinzadeh, B.; Fatikow, S. A novel direct inverse modeling approach for hysteresis compensation of piezoelectric actuator in feedforward applications. *IEEE/ASME Trans. Mechatron.* **2013**, *18*, 981–989. [CrossRef]

21. An, Z.; Xiang, C.; Wang, J.; Dong, L.; Hao, L. Integrated design of micro force sensor oriented to cell micro-operation. *Technol. Health Care* **2015**, *23*, S551–S558. [CrossRef] [PubMed]

22. Kuhnen, K. Modelling, identification and compensation of complex hysteretic nonlinearities—A modified Prandtl-Ishlinskii approach. *Eur. J. Control* **2003**, *9*, 407–418. [CrossRef]

23. Gu, G.-Y.; Yang, M.-J.; Zhu, L.-M. Real-time inverse hysteresis compensation of piezoelectric actuators with a modified Prandtl-Ishlinskii model. *Rev. Sci. Instrum.* **2012**, *83*, 065106. [CrossRef] [PubMed]

micromachines

MDPI

Article

Transparent Ferroelectric Capacitors on Glass

Daniele Sette, Stéphanie Girod, Renaud Leturcq, Sebastjan Glinsek and Emmanuel Defay *

Luxembourg Institute of Science and Technology, Department of Materials Research and Technology,
41 rue du Brill, Belvaux L-4422, Luxembourg; daniele.sette@gmail.com (D.S.); stephanie.girod@list.lu (S.G.);
renaud.leturcq@list.lu (R.L.); sebastjan.glinsek@list.lu (S.G.)
* Correspondence: emmanuel.defay@list.lu; Tel.: +352-275-888-510

Received: 6 September 2017; Accepted: 17 October 2017; Published: 20 October 2017

Abstract: We deposited transparent ferroelectric lead zirconate titanate thin films on fused silica and contacted them via Al-doped zinc oxide (AZO) transparent electrodes with an interdigitated electrode (IDE) design. These layers, together with a TiO_2 buffer layer on the fused silica substrate, are highly transparent (>60% in the visible optical range). Fully crystallized $Pb(Zr_{0.52}Ti_{0.48})O_3$ (PZT) films are dielectrically functional and exhibit a typical ferroelectric polarization loop with a remanent polarization of 15 $\mu C/cm^2$. The permittivity value of 650, obtained with IDE AZO electrodes is equivalent to the one measured with Pt electrodes patterned with the same design, which proves the high quality of the developed transparent structures.

Keywords: transparent piezoelectrics; glass substrate; chemical solution deposition

1. Introduction

In parallel to the development of technologies on silicon, there is a large panel of new functions dedicated to transparent substrates and more specifically to glass [1]. This should involve completely transparent functional layers such as transistors [2] and sensing elements [3]. Transparent pixels or photovoltaic devices have already been published in the literature [4]. Similarly to the More than Moore trend, one can think of developing new devices on glass with optical or mechanical coupling in order to multiply applications. For instance, it has recently been suggested that haptic functions can be added on mobile phone screens through piezoelectric ceramics [5]. It would be ideal to achieve completely transparent stacks on glass so as not to perturb light transmission when it comes to adding new functionalities to screens. We will eventually target transparent piezoelectric stacks in order to fabricate haptic devices. More specifically, we want to study lead zirconate titanate (PZT) films, piezoelectric films with one of the highest reported piezoelectric coefficients, deposited on glass. Moreover, we aim at suppressing the bottom electrode in order to potentially improve the overall transparency. It induces that top electrodes will be interdigitated. Our top choice of electrode is Al-doped zinc oxide (AZO) because of its optical transparency and electrical conductivity [6]. In addition, it does not contain any rare and/or harmful components such as indium in indium tin oxide (ITO) electrodes.

The deposition of PZT on glass substrates, unlike PZT films on silicon, has barely been studied. All examples with electrical results involved a bottom electrode, which was made of Pt or transparent conductive oxides, such as ITO [7] or fluorine-doped tin oxide (FTO) [8]. Khodorov et al. deposited La-doped PZT via the sol–gel method on tin oxide-coated glass substrate [9]. The reflection in the visible range for 130-nm-thick La-doped PZT (PLZT) did not exceed 20%, and no electrical results were performed. Ohno et al. showed that 1.1-μm-thick PZT with compositions close to the morphotropic phase boundary (MPB, corresponding to Zr/Ti = 52/48), deposited via chemical solution deposition (CSD) on soda lime glass and covered with CSD ITO exhibited a dielectric permittivity ε_r close to 1000, a remanent polarization as large as 36 $\mu C/cm^2$, and a longitudinal piezoelectric coefficient d_{33} reaching 120 pC/N after thermal treatment at 600 °C [10]. The transmittance of the stack without top electrode was around 60% in the visible optical range, which was a substantial improvement with respect to Khodorov et al. Liu et al. [11] obtained very similar results on Co-doped PZT also deposited via CSD on an ITO/glass substrate. More recently, Bayraktar et al. [12] focused on the crystalline quality of PZT films grown on glass and reported epitaxial growth of MPB PZT by utilizing $Ca_2Nb_3O_{10}$ and $Ti_{0.87}O_2$ nanosheets as crystalline buffer layers, combined with ITO or Pt/ITO as bottom electrode. Another interesting approach leading to epitaxial films has been proposed by Terada et al. [13], who first grew PZT on MgO and then transferred the layers to glass substrates, though Pt electrodes have been utilized. The unique example of a fully transparent and electrically functional PZT-based stack on glass substrates was published in 2007 by Uprety et al. [14]. The studied stack was ITO/LNO/PZT/LNO/ITO (where LNO stands for $LaNiO_3$) on glass. The PZT was 90-nm-thick, and the overall transparency of the final device capacitor was around 50% before the deposition of the top electrode. In this paper, we propose to improve the overall transparency of the same kind of MPB PZT stack by suppressing the bottom electrode and adopting an in-plane interdigitated electrode (IDE) configuration, similar to what has been performed on silicon more than a decade ago [15].

2. Experimental Section

A 2"-diameter and 500-μm-thick substrate made of optical grade fused silica with roughness Ra <1 nm has been used. A buffer layer was needed between glass and PZT to avoid cracks. Here, this buffer layer was made of 20-nm-thick oxidized titanium deposited via evaporation at 25 °C and annealed at 700 °C in air. Such high temperature infers the complete oxidation of Ti, which stabilizes titanium oxide with respect to the subsequent steps. PZT with MPB composition—meaning Zr/Ti = 52/48—was deposited by spin coating three successive layers of "PZT-E1" commercial sol precursors from MMC (Mitsubishi Materials Corporation, Tokyo, Japan). Each of the three PZT layers was spun, dried at 130 °C for 5 min, and pyrolyzed at 350 °C for 5 min on hot plate in air. A final annealing step performed in a box furnace at 700 °C in air for 30 min induced the crystallization of the 200-nm-thick PZT film in the desired perovskite structure. The 115-nm-thick AZO IDE electrodes were then deposited via atomic layer deposition and patterned through a lift-off process using lift-off resist (LOR) Shipley resist in order to generate an undercut below a resist layer (1813 Shipley). Reference samples made of sputtered 100-nm-thick Pt electrodes with the same IDE design were also realized by standard lithography/ion beam etching process. The absolute values of electric field E, relative dielectric permittivity ε' and polarization P were extracted from the measurements as recently described by Nigon et al. [16]. Because the gap and width of the fingers are significantly larger than the thickness of the film (a, b >> t_f), the effective gap $a + \Delta a$ ($\Delta a = 1.324 \times t_f$) was used in the calculations.

X-ray diffraction patterns were obtained with a Panalytical diffractometer in θ-2θ configuration, - 2θ being the measurement angle of the detector with respect to X-ray incident beam on the sample at angle θ. The light transmittance of the stack was measured on a TECAN (Tecan Group Ltd., Zürich, Switzerland) absorbance instrument. The Agilent atomic force microscope (AFM) was used to measure root-mean-square (RMS) surface roughness. The permittivity-electric field (ε'-E) curves and polarization-electric field (P-E) loops were collected with an Aixacct set-up.

3. Results

Figure 1a shows the X-ray diffraction pattern of crystallized PZT films on glass. The perovskite structure is the only visible phase. No secondary phase such as pyrochlore has been detected. There is no preferred crystalline orientation as one could expect on an amorphous substrate such as fused silica. The RMS value obtained from 5×5 μm^2 areas by AFM is 2.0 nm. Figure 1b shows a top view of the transparent finalized substrate—fused silica/TiO_2/PZT/AZO—as observed with optical microscopy. Note that the logo is not printed on the wafer but is visible through the latter. PZT film is homogeneous and did not encounter cracks during the fabrication process. Our experiments have shown that a buffer layer of TiO$_2$ with a minimum thickness of 10 nm is mandatory to avoid cracks. For thinner layers or no buffer layer at all, PZT crystallizes but exhibits cracks.

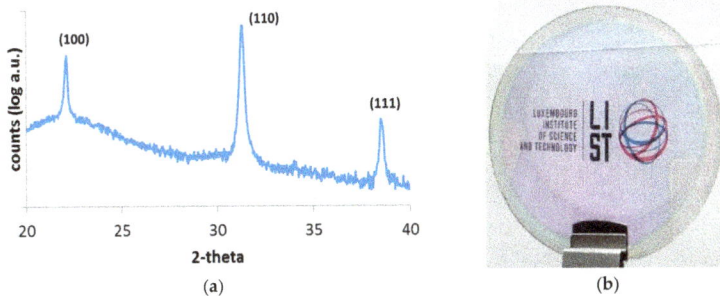

Figure 1. (a) The θ-2θ pattern of 200-nm-thick Pb(Zr$_{0.52}$Ti$_{0.48}$)O$_3$ (PZT) film deposited by spin coating on TiO$_2$/fused silica and crystallized at 700 °C. All peaks correspond to the perovskite structure; (b) Optical image of a 2"-diameter wafer coated with 20 nm of TiO$_2$, 200 nm of PZT, and 115 nm of patterned Al-doped zinc oxide (AZO), the latter being visible in the central area of the wafer. TiO$_2$ exhibits a purplish color. Note that the logo is visible through the glass wafer.

The main drawback of TiO$_2$ is that it induces a purplish color that limits the device's transparency, as can be seen in Figure 1b. Fused silica exhibits transmittance beyond 93% in the visible spectral range (cf. Figure 2). This transmittance decreases down to 65% around $\lambda = 400$ nm after TiO$_2$ has been deposited. Adding PZT creates Fabry–Perot oscillations but does not strongly impact the overall transmittance. It also shifts the minimum wavelength that can cross the wafer from 300 nm with TiO$_2$ to 340 nm if one takes 20% transmittance as a reference. The final 115-nm-thick AZO slightly improves the stack transmittance probably because of the index matching effect. Indeed, the ZnO refractive index is around 2.0, whereas that of PZT is larger, around 2.4. Consequently, capping PZT with ZnO-based material smoothens the gap and therefore acts as an index matching layer. AZO main drawback is that it stops transmittance below 360 nm. All in all, the stack transmittance is higher than 60% in the range from 400 nm to 1000 nm. As mentioned in the introduction, Uprety et al. reported 50% transmittance on their transparent piezoelectric stack involving ITO as the bottom electrode [14]. Therefore, our strategy to suppress the bottom electrode helps in improving transparency.

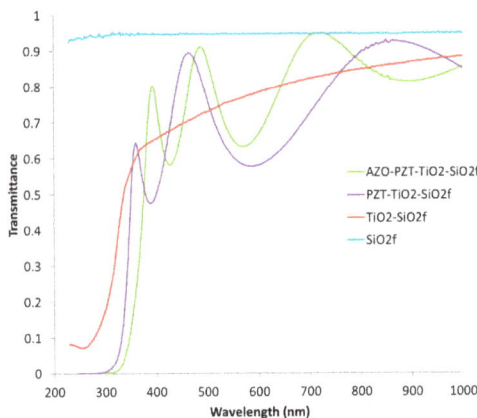

Figure 2. Optical transmission spectra from 232 nm to 1000 nm measured on the successive stacks of AZO/PZT/TiO$_2$/fused silica.

Making electrically functional transparent piezoelectric devices without bottom electrode infers that both electrodes have to be processed on the top of the piezoelectric layer. To do so, we patterned AZO IDE electrodes as shown in Figure 3. Various gaps (5, 10, and 20 μm) were processed. Although AZO is not as conductive as ITO (typical sheet resistance of 250 Ω^2 for 100-nm-thick films in our case), it only involves abundant and low-cost metals (aluminium and zinc).

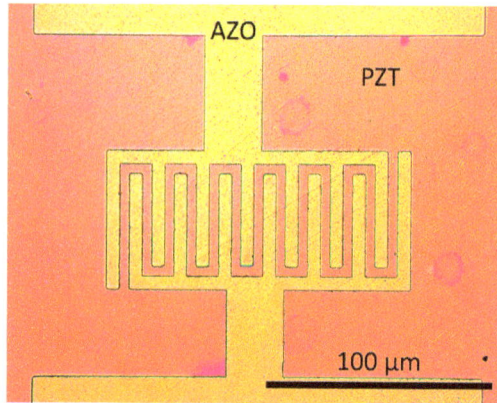

Figure 3. Top-view optical micrograph of a final device showing patterned AZO IDE top electrodes. Here, the gap between subsequent fingers is 5 μm, and each finger is 5-μm-wide.

Figure 4a shows the relative permittivity ε' and dielectric losses tanδ versus the DC electric field of the 7-tooth IDE structure reported in Figure 3. Both curves display the typical hysteretic behavior of ferroelectric films, with respective zero-field values of permittivity and tanδ of 650 and 0.03. Contrary to what is generally observed in metal–insulator–metal structures (MIM), the curves are very symmetrical with respect to the Y-axis. This is a direct consequence of the symmetrical IDE structure that has been processed with the same metal. The losses are very low—below 1%—at voltages as large as 400 kV/cm. It evidences the good quality of PZT, but also the compatibility of AZO electrodes with PZT in this IDE structure.

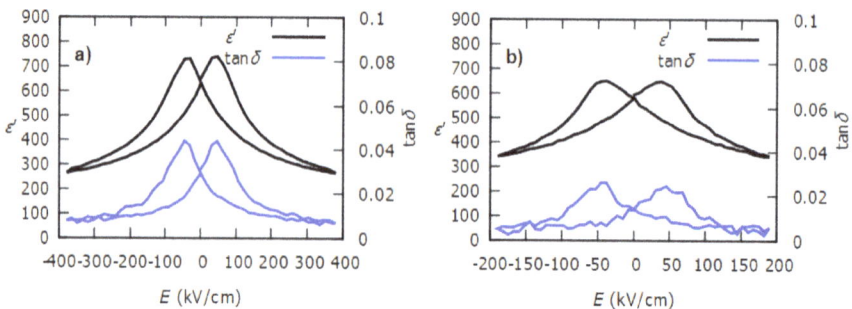

Figure 4. Relative permittivity ε' and tanδ at 1 kHz versus DC electric field E of a 7-tooth IDE structure with a 5 μm gap. Electrodes are made of (**a**) AZO and (**b**) Pt. For permittivity calculation, the stray capacitance was subtracted from the measured capacitance.

Figure 4b represents the same characterization on the same IDE structure, though with Pt electrodes. The difference lies in the maximum electric field applied, which is only 200 kV/cm with Pt. Indeed, this structure was unable to withstand larger field because of the ion milling etching step that weakened PZT and/or induced Pt re-deposition. It ended up with strong extrinsic leakage at a high electric field. In Figure 4, one can spot that permittivity values are very similar for both AZO and Pt electrodes. The latter is considered as the reference metal for PZT. Therefore, it shows that AZO stands for a convincing transparent alternative to Pt in order to contact PZT structures.

Finally, a clear ferroelectric behaviour is observed on the polarization-electric field *(P-E)* loop of PZT on glass with AZO electrodes, as shown in Figure 5. The extracted value of remanent polarization P_r is 15 $\mu C/cm^2$, which is approximately half the value reported for 500-nm-thick (100) MPB PZT on Si with Pt IDE electrodes, but comparable to its counterpart in MIM geometry [17]. The calculated coercive field E_c is 52 kV/cm. The slanted shape of the loop is a measurement artefact due to relatively large stray capacitance compared to the total capacitance value (measured on IDE structures and shown in Figure 3). We wanted to ensure that this tilted loop was not induced by AZO's rather high resistivity, so we compared it with Pt top electrodes patterned with the same design. Both loop shape and coercive voltage are very similar to what has been observed with AZO electrodes. In addition, we found that AZO equivalent resistance of Figure 3 is around 700 Ω. This impedance is negligible compared to PZT, which lies in the GΩ range (C~1 pF) during the *P-E* loop collection performed at 100 Hz. Consequently, AZO's higher resistance has no significant influence on the *P-E* loop's shape.

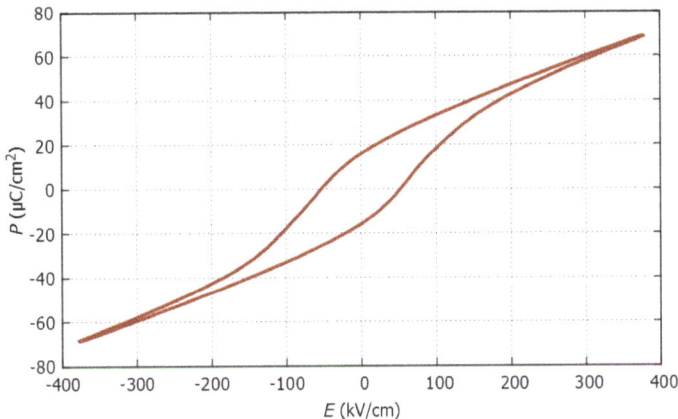

Figure 5. Polarization-electric field *(P-E)* loop of the 200-nm PZT on glass with AZO 5-μm-gap IDE electrodes. The measurement was performed at 100 Hz.

4. Conclusions

In this paper, we have shown that it is possible to make functional transparent ferroelectric structures with PZT and IDE AZO electrodes. The measured permittivity, dielectric losses, and polarization values of 650, 0.03, and 15 $\mu C/cm^2$, respectively, have confirmed typical ferroelectric behavior, which is a pre-requisite of piezoelectric ferroelectric materials such as PZT. The main advantage of the proposed structure is the absence of bottom electrode, which implies improved transparency of the whole PZT stack (>60% in the visible range) if one compares with Uprety et al.'s [14] study with the ITO bottom electrode.

Acknowledgments: The authors thank FNR, the Luxembourg Research National Agency (Fonds National de la Recherche) for granting this study through COFERMAT project (grant number FNR/P12/4853155/Kreisel) belonging to the PEARL scheme. We also thank Patrick Grysan for his help on AFM measurements.

Author Contributions: Daniele Sette and Emmanuel Defay conceived and designed the experiments; Daniele Sette, Stéphanie Girod, and Sebastjan Glinsek performed the experiments; Renaud Leturcq, Sebastjan Glinsek, Daniele Sette, and Emmanuel Defay analyzed the data; Sebastjan Glinsek and Emmanuel Defay wrote the paper.

Conflicts of Interest: The authors declare no conflict of interest.

References

1. Stoppa, M.; Chiolerio, A. Wearable Electronics and Smart Textiles: A Critical Review. *Sensors* **2014**, *14*, 11957–11992. [CrossRef] [PubMed]
2. Fortunato, E.; Barquinha, P.; Martins, R. Oxide semiconductor thin-film transistors: A review of recent advances. *Adv. Mater.* **2012**, *24*, 2945–2986. [CrossRef] [PubMed]
3. Martins, R.; Fortunato, E.; Nunes, P.; Ferreira, I.; Marques, A.; Bender, M.; Katsarakis, N.; Cimalla, V.; Kiriakidis, G. Zinc oxide as an ozone sensor. *J. Appl. Phys.* **2004**, *96*, 1398–1408. [CrossRef]
4. Fortunato, E.; Ginley, D.; Hosono, H.; Paine, D.C. Transparent conducting oxides for photovoltaics. *MRS Bull.* **2007**, *32*, 242–247. [CrossRef]
5. Bernard, F.; Gorisse, M.; Casset, F.; Chappaz, C.; Basrour, S. Design, Fabrication and Characterization of a Tactile Display Based on AlN Transducers. *Procedia Eng.* **2014**, *87*, 1310–1313. [CrossRef]
6. Sette, D.; Girod, S.; Godard, N.; Adjeroud, N.; Chemin, J.B.; Leturcq, R.; Defay, E. Transparent piezoelectric transducers for large area ultrasonic actuators. In Proceedings of the 30th IEEE International Conference on Micro Electro Mechanical Systems (MEMS), Las Vegas, NV, USA, 22–26 January 2017; pp. 793–796.
7. Cotroneo, V.; Davis, W.N.; Marquez, V.; Reid, P.B.; Schwartz, D.A.; Johnson-Wilke, R.L.; Trolier-McKinstry, S.E.; Wilke, R.H. Adjustable grazing incidence X-ray optics based on thin PZT films. In Proceedings of the International Society for Optics and Photonics (SPIE), San Diego, CA, USA, 12–16 August 2012; p. 850309.
8. Pérez-Tomás, A.; Lira-Cantú, M.; Catalan, G. Above-Bandgap Photovoltages in Antiferroelectrics. *Adv. Mater.* **2016**, *28*, 9644–9647. [CrossRef] [PubMed]
9. Khodorov, A.; Gomes, M.J. Preparation and optical characterization of lanthanum modified lead zirconate titanate thin films on indium-doped tin oxide-coated glass substrate. *Thin Solid Films* **2006**, *515*, 1782–1787. [CrossRef]
10. Ohno, T.; Fujimoto, M.; Suzuki, H. Preparation and characterization of PZT thin films on ITO/glass substrate by CSD. *Key Eng. Mater.* **2006**, *301*, 41–44. [CrossRef]
11. Liu, Z.; Liu, Q.; Liu, H.; Yao, K. Electrical properties of undoped PZT and Co-doped PCZT films deposited on ITO/glass substrates by a sol–gel method. *Phys. Stat. Sol. A* **2005**, *202*, 1834–1841. [CrossRef]
12. Bayraktar, M.; Chopra, A.; Bijkerk, F.; Rijnders, G. Nanosheet controlled epitaxial growth of $PbZr_{0.52}Ti_{0.48}O_3$ thin films on glass substrates. *Appl. Phys. Lett.* **2014**, *105*, 132904. [CrossRef]
13. Terada, K.; Suzuki, T.; Kanno, I.; Kotera, H. Hidetoshi Kotera, Fabrication of single crystal PZT thin films on glass substrates. *Vacuum* **2007**, *81*, 571–578. [CrossRef]
14. Uprety, K.K.; Ocola, L.E.; Auciello, O. Growth and characterization of transparent Pb (Zi, Ti)O₃ capacitor on glass substrate. *J. Appl. Phys.* **2007**, *102*, 084107. [CrossRef]
15. Zhang, Q.Q.; Gross, S.J.; Tadigadapa, S.; Jackson, T.N.; Djuth, F.T.; Trolier-McKinstry, S. Lead zirconate titanate films for d33 mode cantilever actuators. *Sens. Actuators A* **2003**, *105*, 91–97. [CrossRef]
16. Nigon, R.; Raeder, T.M.; Muralt, P. Characterization methodology for lead zirconate titanate thin films with interdigitated electrode structures. *J. Appl. Phys.* **2017**, *121*, 204101. [CrossRef]
17. Chidambaram, N.; Balma, D.; Nigon, R.; Mazzalai, A.; Matloub, R.; Sandu, C.S.; Muralt, P. Converse mode piezoelectric coefficient for lead zirconate titanate thin film with interdigitated electrode. *J. Micromech. Microeng.* **2015**, *25*, 045016. [CrossRef]

micromachines

MDPI

Article

Design and Simulation of A Novel Piezoelectric AlN-Si Cantilever Gyroscope

Jian Yang [1,2,3], Chaowei Si [1,*], Fan Yang [1,2], Guowei Han [1], Jin Ning [1,3,4], Fuhua Yang [1,2] and Xiaodong Wang [1,5,*]

1   Institute of Semiconductors, Chinese Academy of Sciences, Beijing 100083, China;
    yangjian@semi.ac.cn (J.Y.); yangfan3104@semi.ac.cn (F.Y.); hangw1984@semi.ac.cn (G.H.);
    ningjin@semi.ac.cn (J.N.); fhyang@semi.ac.cn (F.Y.)
2   College of Materials Science and Opto-Electronic Technology, University of Chinese Academy of Sciences,
    Beijing 100049, China
3   State Key Laboratory of Transducer Technology, Chinese Academy of Sciences, Beijing 100083, China
4   School of Electronic, Electrical and Communication Engineering, University of Chinese Academy of
    Sciences, Beijing 100049, China
5   School of Microelectronics, University of Chinese Academy of Sciences, Beijing 100049, China
*   Correspondences: schw@semi.ac.cn (C.S.); xdwang@semi.ac.cn (X.W.); Tel.: +86-10-8230-5147 (C.S.);
    +86-10-8230-5042 (X.W.)

Received: 28 December 2017; Accepted: 13 February 2018; Published: 15 February 2018

Abstract: A novel design of piezoelectric aluminum nitride (AlN)-Si composite cantilever gyroscope is proposed in this paper. The cantilever is stimulated to oscillate in plane by two inverse voltages which are applied on the two paralleled drive electrodes, respectively. The whole working principles are deduced, which based on the piezoelectric equation and elastic vibration equation. In this work, a cantilever gyroscope has been simulated and optimized by COMSOL Multiphysics 5.2a. The drive mode frequency is 87.422 kHz, and the sense mode frequency is 87.414 kHz. The theoretical sensitivity of this gyroscope is 0.145 pm/○/s. This gyroscope has a small size and simple structure. It will be a better choice for the consumer electronics.

Keywords: microelectromechanical systems (MEMS); aluminum nitride; piezoelectric effect; cantilever; gyroscope; simulation

---

1. Introduction

Microelectromechanical systems (MEMS) gyroscopes are core inertial sensors. They are widely used in smartphones, unmanned aerial vehicles (UAV), automotive electronics, or other consumer goods [1]. In recent years, aluminum nitride (AlN) piezoelectric gyroscopes become a new research trend. The AlN gyroscopes can be categorized into two main types: (1) thin-film AlN gyroscopes and (2) bulk acoustic wave (BAW) AlN-on-Si gyroscopes. The thin-film AlN gyroscopes have traditional structures: beams and proof masses. The AlN thin-film gyroscope relies on bending vibration of beams. The drive mode is simulated by voltages and the Coriolis force is detected through piezoelectric charge. While, the BAW AlN-on-Si gyroscopes are made up of an AlN film on silicon substrate with a specific geometry. Two degenerate bulk acoustic wave modes of the AlN film, which are orthogonal to each other, are used as the drive mode and sense mode, respectively.

The Albert P. Pisano's group, which comes from the University of California, focuses on the thin-film AlN gyroscopes [2–4]. The structure contains three layers: the AlN layer and the bottom/top electrode layers. The thickness of AlN is 2 μm. These gyroscopes have a high efficiency of electromechanical transduction. However, a near zero stress AlN film is needed. It is difficult for the AlN deposition process, and the robustness of this gyroscope is another serious problem.

Farrokh Ayazi's group, which is from the Georgia Institute of Technology, focuses on the BAW AlN-on-Si gyroscopes [5–7]. From 2013 to the present, many different kinds of BAW gyroscopes have been researched and optimized. These gyroscopes have a relatively high frequency (3.1–11 MHz) and perfect immunity to shock and vibration. However, a high frequency will lead to a reduction of response amplitude. The sensitivity will decrease. Besides, the size of these BAW gyroscopes are very large.

Considering the characteristics of the two AlN gyroscopes, a novel AlN-Si composite cantilever gyroscope is proposed in this paper. Based on the piezoelectric effect of AlN, the cantilever will be excited to bend in-plane. The working frequency is 87 kHz, which is lower than the frequency of BAW AlN-on-Si gyroscopes. This will benefit the sensitivity. This gyroscope has a small size and simple structure. Analysis and simulation of this piezoelectric gyroscope has been carried out and reported on in this paper.

2. Working Principles

2.1. Working Principles of Drive Mode

The schematic of AlN-Si composite cantilever gyroscope is shown in Figure 1. It contains four layers. The bottom electrode layer is Mo, connecting the ground in electric field. The top layer is Al, divided into five parts. The five electrodes include two drive electrodes, two drive-detected electrodes, and one sense electrode. In this gyroscope, AlN layer works as drive function for the composite cantilever. Inverse voltages are applied on two drive electrodes respectively, as shown in Figures 1 and 2. The drive-detected electrodes are used to detect the drive signal. The purpose is to stabilize both magnitude and frequency of the drive signal. Based on the piezoelectric effect of AlN, the two inverse voltages will excite two inverse stresses ($\pm\sigma$)—the compressive stress and tensile stress. The direction of the stresses are along with the length of cantilever. They will form a couple stress. The formula of σ has been deduced as shown in Equation (1).

$$\sigma = E_{AlN}\frac{U}{d_{AlN}}d_{31} = \frac{E_{AlN}U_0\sin(\omega t)d_{31}}{d_{AlN}}$$
$$U = U_0\sin(\omega t)$$

$$(1)$$

where E_{AlN} is the Young's modulus of AlN, d_{AlN} is the thickness of AlN, U is the sinusoidal drive voltage, ω is the angular frequency of drive voltage, and d_{31} is the piezoelectric coefficient of AlN. According to the references, $d_{31} = -2.6$ pm/V [8,9].

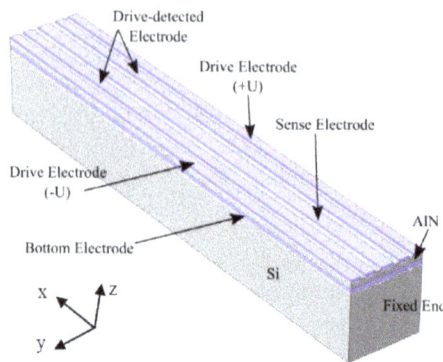

Figure 1. The structure of aluminum nitride (AlN)-Si composite cantilever gyroscope.

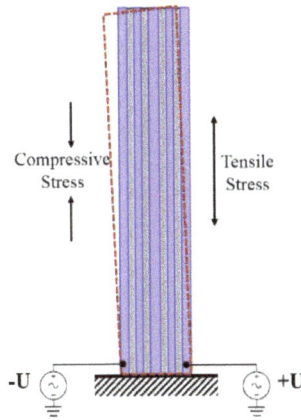

Figure 2. The schematic of in-plane vibration of cantilever.

This couple stress will drive the AlN-Si composite cantilever to bend in-plane, as shown in Figure 2. The bending moment M is given by Equation (2). Where W represents the width of cantilever and W_{dri} represents the width of drive electrodes.

$$M = \sigma W_{dri} d_{AlN}(W - W_{dri})$$
$$= E_{AlN} U_0 \sin(\omega t) d_{31} W_{dri}(W - W_{dri}) \tag{2}$$

The deformation of cantilever satisfies the approximately differential equation of flexural curve of Equation (3). By solving the Equation (3), the displacement function $y(x,t)$ can be deduced as shown in Equation (4). Where EI_y denotes the flexural rigidity of the composite cantilever in y direction. E_{Al}, E_{Mo}, and E_{Si} are the Young's modulus of Al, Mo, and Si respectively; and d_{Al}, d_{Mo}, and d_{Si} are the thickness of Al, Mo, and Si respectively. The x axial is parallel to the length direction of cantilever. x is the position, belonging to [0,L]. All the parameters are shown in Table 1.

$$y'' = \frac{M}{EI_y} \tag{3}$$

$$y(x,t) = \frac{6E_{AlN} d_{31} W_{dri}(W - W_{dri})}{W^3(E_{Al}d_{Al} + E_{AlN}d_{AlN} + E_{Mo}d_{Mo} + E_{Si}d_{Si})} x^2 U_0 \sin(\omega t) \tag{4}$$

Table 1. Parameters of aluminum nitride (AlN)-Si composite cantilever for calculation and simulation.

Names	Parameters	Values	Units
E_{AlN}	Young's modulus	410	GPa
E_{Si}	Young's modulus	160	GPa
E_{Mo}	Young's modulus	312	GPa
E_{Al}	Young's modulus	70	GPa
d_{AlN}	Thickness	1.5	μm
d_{Si}	Thickness	20	μm
d_{Mo}	Thickness	0.3	μm
d_{Al}	Thickness	0.3	μm
d_{31}	Piezoelectric coefficient	−2.6	pm/V
L	Length of cantilever	600	μm
W	Width of cantilever	22.87	μm
W_{dri}	Width of drive electrode	3	μm
U_0	Amplitude	1	V

Plugging the values into Equation (4), the theoretical value of maximum displacement $y_{max} = y(L, \frac{2n+1}{2\omega}\pi) = 2.920$ nm can be calculated. Meanwhile, the same cantilever model has been designed and simulated by COMSOL Multiphysics 5.2a (COMSOL Inc., Stockholm, Sweden). A stationary study has been simulated. The maximum stationary displacement of simulation value is $y_{s\text{-max}} = 2.878$ nm, as shown in Figure 3. Comparing the theoretical value to the simulation one, an error ratio 1.5% can be calculated. This proves that the deduction of working principles is logical and accurate. To further research the resonant properties, a frequency domain study has been done. The maximum resonant displacement is 9.92 µm, and the quality factor is 3450, as shown in Figure 4.

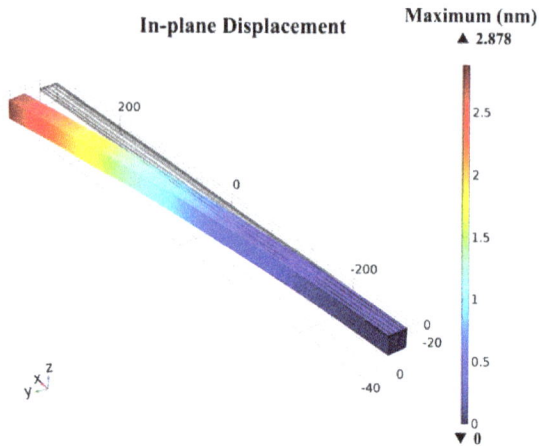

Figure 3. Displacement simulation of AlN-Si composite cantilever under ±1 V drive voltages, stationary study.

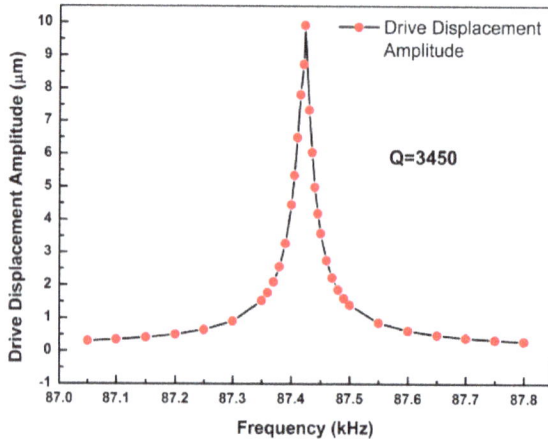

Figure 4. The resonant displacement amplitude of drive mode, frequency domain study.

2.2. Working Principles of Sense Mode

This gyroscope can detect an angular rate which is along the length direction of cantilever. When an x axial angular rate Ω is applied, the cantilever will be driven to bend out-plane by Coriolis force $F_C(x,t)$. The $F_C(x,t)$ formula can be deduced as shown in Equation (5). The ρ_{all} is a linear density of the composite cantilever. C_0 represents the constant coefficient of $y(x,t)$. The direction of $F_C(x,t)$ is z-axial.

$$F_C(x,t) = 2m(x)\Omega \dot{y}(x,t)$$
$$= \int_0^x 2\rho_{all}\Omega \dot{y}(x_0,t)dx_0 \qquad (5)$$
$$= \frac{2}{3}\rho_{all}\Omega \omega C_0 \cos(\omega t)x^3$$

$$\rho_{all} = \frac{\rho_{Al}V_{Al} + \rho_{AlN}V_{AlN} + \rho_{Mo}V_{Mo} + \rho_{Si}V_{Si}}{L}$$

$$C_0 = \frac{6E_{AlN}d_{31}W_{dri}(W - W_{dri})}{W^3(E_{Al}d_{Al} + E_{AlN}d_{AlN} + E_{Mo}d_{Mo} + E_{Si}d_{Si})}U_0$$

By solving the approximately differential equation of flexural curve in z direction-Equation (6), a displacement function $z(x,t)$ can be deduced. The function $z(x,t)$ is shown in Equation (7). EI_z is the flexural rigidity of the composite cantilever in z direction. The function $z(x,t)$ is the displacement of sense mode. Therefore, a linear relation between z and Ω can be gotten from Equation (7). The maximum coefficient of $z(\Omega)$ is the sensitivity of this gyroscope.

$$z'' = \frac{M_z(x,t)}{EI_z} \qquad (6)$$

$$M_z(x,t) = \int_x^L F_C(x_0,t)\cdot(L - x_0)dx_0$$

$$z(x,t) = \frac{\rho_{all}\omega C_0\Omega}{15EI_z}(\frac{1}{4}L^5x^2 - \frac{1}{12}Lx^6 + \frac{1}{21}x^7)\cos(\omega t) \qquad (7)$$

3. Design of AlN-Si Composite Cantilever Gyroscope

The main structure of this gyroscope is shown in Figure 1. The sizes of AlN-Si composite cantilever are designed and optimized by COMSOL Multiphysics. To realize the mode matching, the parametric sweep function was used. The valve of the cantilever width has been swept from 20 μm to 25 μm, with a step of 0.01 μm. When the value is 22.87 μm, the frequencies of drive mode and sense mode are nearly equal. All the values are shown in Table 2. The simulation result of drive mode frequency is 87.422 kHz, and the sense mode frequency is 87.414 kHz. The mode shapes are shown in Figures 5 and 6, respectively.

Table 2. Sizes of AlN-Si composite cantilever gyroscope.

Names	Parameters	Values	Units
L	Length of cantilever	600	μm
W	Width of cantilever	22.87	μm
W_{dri}	Width of drive electrode	3	μm
W_{dri_det}	Width of drive-detected electrode	3	μm
W_{sen}	Width of sense electrode	4	μm
d_{AlN}	Thickness	1.5	μm
d_{Si}	Thickness	20	μm
d_{Mo}	Thickness	0.3	μm
d_{Al}	Thickness	0.3	μm

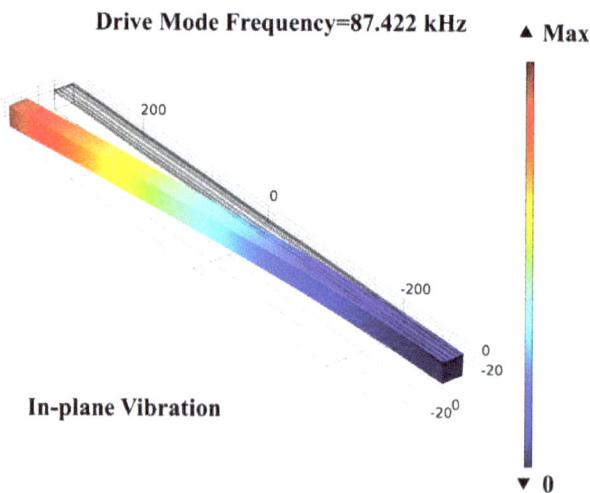

Figure 5. The drive mode simulation of the cantilever gyroscope.

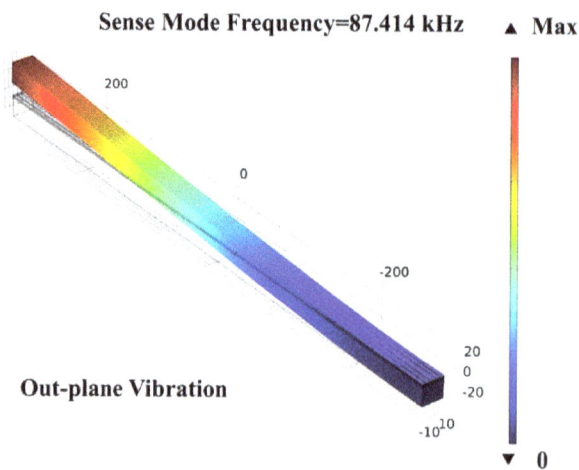

Figure 6. The sense mode simulation of the cantilever gyroscope.

The quality factors of this gyroscope are assumed to be $Q_{drive} = 3000$ and $Q_{sense} = 3000$, respectively. The maximum resonant displacement $z_{res}(L, \frac{n\pi}{\omega})$ is shown in Equation (8). Therefore, the sensitivity of this gyroscope is 0.145 pm/o/s, as shown in Figure 7.

$$z_{res}\left(L, \frac{n\pi}{\omega}\right) = Q_{drive}Q_{sense}z\left(L, \frac{n\pi}{\omega}\right) = 1.45 \times 10^{-13} \, \Omega \tag{8}$$

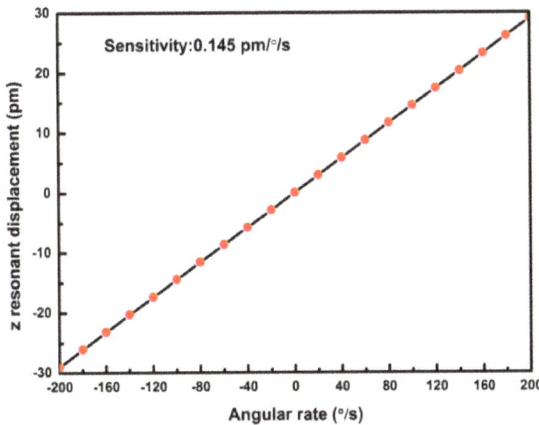

Figure 7. Theoretical sensitivity of the gyroscope—the resonant displacement of z-axis vs. angular rate Ω.

4. Conclusions

The AlN-Si composite cantilever gyroscope is based on the novel electrode design and working principles. Because of the piezoelectric effect of AlN, an in-plane movement has been stimulated by two inverse voltages. The AlN-Si composite cantilever gyroscope has been designed with 87.422 kHz drive frequency and 87.414 kHz sense frequency. The mode-matching has been realized by optimizing the width of the cantilever. This gyroscope has a small size, simple structure, and lower requirements for the processing of AlN film. Hence, this cantilever gyroscope shows great potential in the piezoelectric research and consumer electronics fields.

Acknowledgments: The authors gratefully acknowledge Chinese National Science Foundation (contracts No. 61504130, no. 61704165, and no. 61474115).

Author Contributions: Jian Yang and Chaowei Si conceived and designed the MEMS gyroscope structure. Jian Yang performed the simulation and wrote the paper. Fan Yang, Guowei Han, Jin Ning, Fuhua Yang, and Xiaodong Wang revised the paper.

Conflicts of Interest: The authors declare no conflict of interest.

References

1. Sara, S.; Naser, E. Activity recognition using fusion of low-cost sensors on a smartphone for mobile navigation application. *Micromachines* **2015**, *6*, 1100–1134. [CrossRef]
2. Gabriele, V. MEMS Aluminum Nitride Technology for Inertial Sensors. Doctoral Thesis, University of California, Berkeley, CA, USA, 2011.
3. Fabian, T.G.; Gabriele, V.; Igor, I.I. Novel thin-film piezoelectric aluminum nitride rate gyroscope. In Proceedings of the 2012 IEEE International Ultrasonics Symposium, Dresden, Germany, 7–10 October 2012; pp. 1067–1070.
4. Fabian, G.; Kirti, M.; Kansho, Y. Experimentally validated aluminum nitride based pressure, temperature and 3-axis acceleration sensors integrated on a single chip. In Proceedings of the 2014 IEEE 27th International Conference on Micro Electro Mechanical Systems, San Francisco, CA, USA, 26–30 January 2014; pp. 729–732.
5. Roozbeh, T.; Mojtaba, H.S.; Farrokh, A. High-frequency AlN-on-silicon resonant square gyroscopes. *J. Microelectromech. Syst.* **2013**, *22*, 1007–1009.
6. Mojtaba, H.S.; Arashk, N.S.; Roozbeh, T. A dynamically mode-matched piezoelectrically transduced high-frequency flexural disk gyroscope. In Proceedings of the 2015 28th IEEE International Conference on Micro Electro Mechanical Systems (MEMS), Estoril, Portugal, 18–22 January 2015; pp. 789–792.

7. Mojtaba, H.S.; Arashk, N.S.; Farrokh, A. Eigenmode operation as a quadrature error cancellation technique for piezoelectric resonant gyroscopes. In Proceedings of the 2017 IEEE 30th International Conference on Micro Electro Mechanical Systems (MEMS), Las Vegas, NV, USA, 22–26 January 2017; pp. 1107–1110.
8. Hernando, J.; Sanchez-Rojas, J.L.; Gonzalez-Castilla, S. Simulation and laser vibrometry characterization of piezoelectric AIN thin films. *J. Appl. Phys.* **2008**, *104*, 053502-1–053502-9. [CrossRef]
9. Martin, F.; Muralt, P.; Dubois, M.A. Thickness dependence of the properties of highly c-axis textured AlN thin films. *J. Vac. Sci. Technol. A Vac. Surf. Films* **2004**, *22*, 361–365. [CrossRef]

micromachines

MDPI

Article

Development of Piezo-Driven Compliant Bridge Mechanisms: General Analytical Equations and Optimization of Displacement Amplification

Huaxian Wei [1,2], **Bijan Shirinzadeh** [2], **Wei Li** [1,*], **Leon Clark** [2], **Joshua Pinskier** [2] **and Yuqiao Wang** [1]

[1] School of Mechatronic Engineering, China University of Mining and Technology, Xuzhou 221116, China; weihuaxian@yahoo.com (H.W.); cumtwyq@126.com (Y.W.)
[2] Robotics and Mechatronics Research Laboratory, Department of Mechanical and Aerospace Engineering, Monash University, Clayton 3800, Australia; bijan.shirinzadeh@monash.edu (B.S.); leon.s.clark@gmail.com (L.C.); joshua.pinskier@monash.edu (J.P.)
* Correspondence: liweicumt@cumt.edu.cn; Tel./Fax: +86-516-8388-5829

Received: 12 July 2017; Accepted: 27 July 2017; Published: 3 August 2017

Abstract: Compliant bridge mechanisms are frequently utilized to scale micrometer order motions of piezoelectric actuators to levels suitable for desired applications. Analytical equations have previously been specifically developed for two configurations of bridge mechanisms: parallel and rhombic type. Based on elastic beam theory, a kinematic analysis of compliant bridge mechanisms in general configurations is presented. General equations of input displacement, output displacement, displacement amplification, input stiffness, output stiffness and stress are presented. Using the established equations, a piezo-driven compliant bridge mechanism has been optimized to maximize displacement amplification. The presented equations were verified using both computational finite element analysis and through experimentation. Finally, comparison with previous studies further validates the versatility and accuracy of the proposed models. The formulations of the new analytical method are simplified and efficient, which help to achieve sufficient estimation and optimization of compliant bridge mechanisms for nano-positioning systems.

Keywords: flexure hinge; compliant bridge mechanisms; micro-motion scaling; kinematics

1. Introduction

In recent decades, piezoelectric actuators (PZTs) have been frequently used in micro/nano-applications including advanced manufacturing, high precision positioning, scanning probe microscopes and biological cell manipulation [1–4]. The advantages of piezoelectric actuators include precise motion capability, compact size and large blocking force. However, one of their main drawbacks is the relatively small motion stroke, at about 0.1 percent of its length. Consequently, compliant mechanisms are generally employed to scale the displacement in values compatible with PZTs, including bridge [5], Scott-Russell [6], and lever type mechanisms [7].The compliant mechanisms employ flexure hinges instead of rigid joints to eliminate mechanical play and friction, and hence can achieve ultra-precise and smooth motions [8,9]. However, the kinematics of these flexure-based mechanisms is based on the deflections of their flexure hinges, and this has led to techniques for design, analysis and modeling for compliant mechanisms [10–12].

Among the commonly used micro-motion scaling mechanisms, the compliant bridge mechanisms, as shown in Figure 1, have been widely used because of their symmetry, compactness and large magnification capability. In the last decade, compliant bridge mechanisms have been widely employed in flexure-based micro-manipulators to provide amplified piezo-actuations [13,14]. With the increasing demands for high-dexterity manipulation, compliant bridge mechanisms have been used as a regular model to construct more complex structures with multi-degrees of freedom [15]. This has led to the requirement for developing an efficient analytical model of displacement amplification for compliant bridge mechanisms.

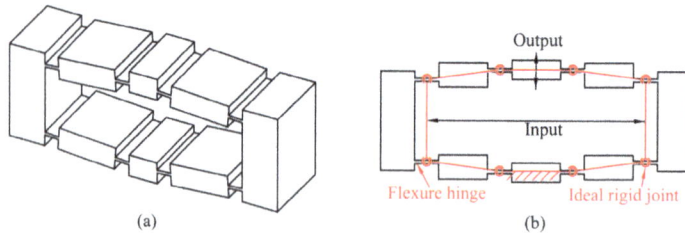

Figure 1. The compliant bridge mechanisms: (**a**) three-dimensional model; (**b**) ideal kinematic model.

Much research has been directed towards deducing analytical models for compliant bridge mechanisms. Ideal kinematic methods, which treat the flexure hinges as ideal revolute joints, have been shown to be inaccurate, owing to their neglecting elastic deformations in flexure hinges [16,17]. Therefore, an analytical model based on Castigliano's displacement theory has been developed by Lobontiu [18]. In addition, the matrix method has also been employed as simplified finite element analysis (FEA) [19]. However, the cumbersome formulations of these methods have limited their application. Methods based on elastic beam theory and motion analyses have been used, where analytical equations of displacement amplification and stiffness are obtained [20]. In addition, non-linear models incorporating beam theory of the flexure hinge for high frequencies or large deformation have been developed [21,22]. However, these studies have focused on the analyses of compliant bridge mechanisms that are specifically in parallel [23], aligned [24] and rhombic type [25,26] configurations, as shown in Figure 2. As a result, design processes are separated and repeated for these configurations since the geometric characteristics are not transformable [27,28]. In addition, the design of a compliant bridge mechanism is simultaneously limited by kinematics, stress and stiffness, which are determined by the geometric parameters. Unlike traditional rigid joints, the orientation of the flexure hinge has a significant influence on the mechanism's performance [29]. For a given application, the optimal design may occur in any of the aforementioned configurations, and hence generalized analytical equations are required for design searches.

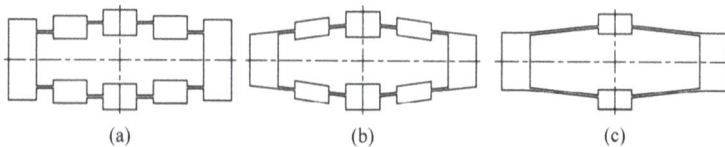

Figure 2. Three types of compliant bridge mechanisms: (**a**) parallel; (**b**) aligned and (**c**) rhombic.

The aim in this paper is to investigate a simplified analytical model to be employed within the optimization of displacement amplification for compliant bridge mechanisms covering all types of configuration. In the following section, a method based on beam theory and kinematic analysis is detailed, and analytical equations of input, output, displacement amplification, stiffness and stress are formulated. Subsequently, optimal designs of piezo-driven compliant bridge mechanisms in terms of displacement amplification under kinematic, stress and stiffness constraints have been established. The presented models and optimizations are then verified by FEA and experimental tests. Finally, comparisons of the established models with previous models are carried out, and a theoretic displacement amplification ratio formula of aligned-type compliant bridge mechanisms is attained.

2. The General Analytical Model

As compliant bridge mechanisms generally employ quadrilateral symmetric structures, a general quarter model of the mechanism is analyzed, as shown in Figure 3. The model is composed of five parts: input link *a*, flexure hinge *b*, middle link *c*, flexure hinge *d* and output link *e*. For simplification, four nodes numbered from 2 to 5 are identified between the conjunctions of each part. Six geometric parameters, which are henceforth called *configuration parameters*, are sufficient to determine the configuration of the general compliant bridge mechanisms, as shown in Figure 3a, namely the lengths and orientations of the two flexure hinges and the middle link (l_2, l_3, l_4, δ_2, δ_3, and δ_4). Without loss of generality, the positive directions of orientation angles are defined as shown in Figure 3a, when the central axes of these parts rotate in anticlockwise direction from horizontal position.

The operation can be illustrated by means of the quarter model, as shown in Figure 3b. From the point of view of the mechanics of materials, the flexure hinges deform under the driving forces (F_X) from the PZT on the input link and the manipulating force (F_Y) on the output link, and this results in a translational input displacement (X_{in}) and a translational output displacement (Y_{out}) due to the symmetric constraints. The positive directions of the input and output forces and displacements are defined as shown in Figure 3b.

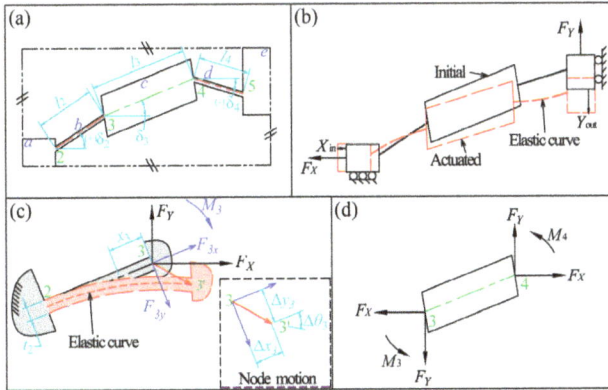

Figure 3. Analytical model of the compliant bridge mechanism: (**a**) general quarter model with configuration parameters; (**b**) schematic of working status; (**c**) deformations of flexure hinge *b*; and (**d**) force equilibrium of the middle link.

2.1. Input and Output Analyses

In order to determine the input and output motions of the compliant bridge mechanism, deflection analyses of flexure hinges are required. Firstly, flexure hinges are analyzed as cantilever beams. Consider the flexure hinge *b*, as shown in Figure 3c, for example, freeing the end (node 3) that is connected to the middle link and let the other end (node 2) be fixed. Using beam theory, the deflections and loads on flexure hinge can be analyzed according to its compliances, that is:

$$\begin{cases} \Delta x_3 = c_{11}^b \cdot F_{3x} \\ \Delta y_3 = c_{22}^b \cdot F_{3y} + c_{23}^b \cdot M_3 \\ \Delta \theta_3 = c_{32}^b \cdot F_{3y} + c_{33}^b \cdot M_3 \end{cases} \tag{1}$$

where Δx_3, Δy_3 and $\Delta \theta_3$ are the axial deformation, deflection and slope angle of flexure hinge *b* at node 3, respectively. F_{3x}, F_{3y} and M_3 are the axial force, shear force and bending moment, respectively. *c* is the compliance factor of the flexure hinge which is solely determined by the geometric parameters and material characteristics. For strip-type flexure hinges, the compliances are given as [30]:

$$\begin{cases} c_{11}^b = \dfrac{l_2}{Ewt_2^3} \\[4pt] c_{22}^b = \dfrac{4l_2^3}{Ewt_2^3} + \dfrac{l_2}{Gwt_2} \\[4pt] c_{23}^b = c_{32}^b = \dfrac{6l_2^2}{Ewt_2^3} \\[4pt] c_{33}^b = \dfrac{12l_2}{Ewt_2^3} \end{cases} \tag{2}$$

where t_2 is the thickness of flexure hinge, w the width of the mechanism, E the modulus of elasticity, and G the modulus of shear. The axial and shear forces on the free end can be obtained by means of force equilibrium of the mechanism, which can be written as:

$$\begin{cases} F_{3x} = F_X \cdot \cos \delta_2 + F_Y \cdot \sin \delta_2 \\ F_{3y} = F_X \cdot \sin \delta_2 - F_Y \cdot \cos \delta_2 \end{cases} \tag{3}$$

where F_{3x} and F_{3y} are the axial and deflecting forces of flexure hinge b at node 3, respectively. Similarly, the axial and shear forces of flexure hinge d can be obtained as:

$$\begin{cases} F_{4x} = F_X \cdot \cos \delta_4 + F_Y \cdot \sin \delta_4 \\ F_{4y} = F_X \cdot \sin \delta_4 - F_Y \cdot \cos \delta_4 \end{cases} \tag{4}$$

The motion of flexure hinge d at node 4 can similarly be identified as: Δx_4, Δy_4 and $\Delta \theta_4$. Equations (3) and (4) indicate that the internal loads, and hence the bending moments, within the two flexure hinges are different if they have different orientations. Since the middle link is treated as rigid, the slope angles of the two flexure hinges at node 3 and 4 are always identical. Considering the force equilibrium of the middle link as shown in Figure 3d, an equation system can be established that relates the bending moments of the two flexure hinges, and can be written as:

$$\begin{cases} F_{3y} \cdot c_{32}^b + M_3 \cdot c_{33}^b = F_{4y} \cdot c_{32}^d + M_4 \cdot c_{33}^d \\ F_X \cdot l_3 \cdot \sin \delta_3 = M_3 + M_4 + F_Y \cdot l_3 \cdot \cos \delta_3 \end{cases} \tag{5}$$

where M_3 and M_4 are the bending moments at node 3 and 4, respectively. By substituting Equations (1)–(4) into Equation (5), the bending moments can be deduced as:

$$\begin{cases} M_3 = \dfrac{F_Y \cdot c_{32}^b \cdot \cos \delta_2 - F_Y \cdot c_{32}^d \cdot \cos \delta_4 - F_X \cdot c_{32}^b \cdot \sin \delta_2 + F_X \cdot c_{32}^d \cdot \sin \delta_4 - F_Y \cdot c_{33}^d \cdot l_3 \cdot \cos \delta_3 + F_X \cdot c_{33}^d \cdot l_3 \cdot \sin \delta_3}{c_{33}^b + c_{33}^d} \\[8pt] M_4 = \dfrac{F_Y \cdot c_{32}^d \cdot \cos \delta_4 - F_Y \cdot c_{32}^b \cdot \cos \delta_2 + F_X \cdot c_{32}^b \cdot \sin \delta_2 - F_X \cdot c_{32}^d \cdot \sin \delta_4 - F_Y \cdot c_{33}^b \cdot l_3 \cdot \cos \delta_3 + F_X \cdot c_{33}^b \cdot l_3 \cdot \sin \delta_3}{c_{33}^b + c_{33}^d} \end{cases} \tag{6}$$

Eventually, the translational displacements of input and output links are composed of deflections of the two flexure hinges and the arc motion of the middle link, which can be written as:

$$\begin{cases} X_{in} = \Delta x_3 \cdot \cos \delta_2 + \Delta y_3 \cdot \sin \delta_2 + \Delta x_4 \cdot \cos \delta_4 + \Delta y_4 \cdot \sin \delta_4 + \Delta \theta_3 \cdot l_3 \cdot \sin \delta_3 \\ Y_{out} = \Delta y_3 \cdot \cos \delta_2 - \Delta x_3 \cdot \sin \delta_2 + \Delta y_4 \cdot \cos \delta_4 - \Delta x_4 \cdot \sin \delta_4 + \Delta \theta_3 \cdot l_3 \cdot \cos \delta_3 \end{cases} \tag{7}$$

By substituting Equation (1) into Equation (7), the closed-form equations of the input and output displacements can be deduced in the form:

$$\begin{cases} X_{in} = a_{11} \cdot F_X + a_{12} \cdot F_Y \\ Y_{out} = a_{21} \cdot F_X + a_{22} \cdot F_Y \end{cases} \tag{8}$$

where $a_{11} - a_{22}$ are coefficients determined by geometric parameters and material characteristics as detailed in Appendix A. Based on the equation system, the analytical equations of displacement amplification, input and output stiffness can be deduced with simplified formulations.

2.2. Displacement Amplification

The displacement amplification is the ratio of the output displacement to the input displacement when the output link is free. Referring to Equation (8), the displacement amplification can be deduced as:

$$da = \frac{a_{21}}{a_{11}} = \frac{\cos\delta_2\cdot\left(c_{22}^b\cdot\sin\delta_2 - c_{11}^b\cdot\sin\delta_2 + B\right) + \cos\delta_4\cdot\left(c_{22}^d\cdot\sin\delta_4 - c_{11}^d\cdot\sin\delta_4 + A\right) + \frac{l_3\cdot\cos\delta_3\cdot C}{c_{33}^b + c_{33}^d}}{\sin\delta_2\cdot\left(c_{22}^b\cdot\sin\delta_2 + B\right) + \sin\delta_4\cdot\left(c_{22}^d\cdot\sin\delta_4 + A\right) + c_{11}^b\cdot\cos^2\delta_2 + c_{11}^d\cdot\cos^2\delta_4 + \frac{l_3\cdot\sin\delta_3\cdot C}{c_{33}^b + c_{33}^d}} \tag{9}$$

in which $A = \dfrac{c_{23}^d\cdot\left(c_{32}^b\cdot\sin\delta_2 - c_{32}^d\cdot\sin\delta_4 + c_{33}^b\cdot l_3\cdot\sin\delta_3\right)}{c_{33}^b + c_{33}^d}$, $B = \dfrac{c_{23}^b\cdot\left(c_{32}^d\cdot\sin\delta_4 - c_{32}^b\cdot\sin\delta_2 + c_{33}^d\cdot l_3\cdot\sin\delta_3\right)}{c_{33}^b + c_{33}^d}$,
$C = c_{32}^b\cdot c_{33}^d\cdot\sin\delta_2 + c_{32}^d\cdot c_{33}^b\cdot\sin\delta_4 + c_{33}^b\cdot c_{33}^d\cdot l_3\cdot\sin\delta_3$.

2.3. Input and Output Stiffness

The input stiffness of the compliant bridge mechanism is defined as the applied input force corresponding to unit input displacement, whilst the output link is free. Similarly, an equation system can be found as:

$$k_{in} = \frac{F_X}{X_{in}} = \frac{1}{a_{11}} \tag{10}$$

In addition, the output stiffness of the compliant bridge mechanism is defined as the applied output force per unit output of displacement when the input link is free. Consequently, an equation system can be established for the output stiffness:

$$k_{out} = \frac{F_Y}{Y_{out}} = \frac{1}{a_{22}} \tag{11}$$

2.4. Stress Analysis

For compliant mechanisms, the maximum motion range is also limited by the maximum stress in the structure. The maximum stress is generated under the maximum loads. Since the positive output force tends to decrease the stress in the flexure hinge, only input force on the input link is taken into consideration, which can be written as:

$$F_X^{max} = F_{PZT}^{max} + F_{preload} \tag{12}$$

where $F_{preload}$ is the preload which is usually essential to eliminate clearance between PZT and the structure. F_{PZT}^{max} is the maximum actuating force from the PZT corresponding to the maximum input displacement, by referring to Equation (10), which can be written as:

$$F_{PZT}^{max} == X_{PZT}^{nl}\cdot k_{in} \tag{13}$$

where X_{PZT}^{nl} is the nominal stroke of the PZT. In addition, the true strokes of PTZs are reduced by the compression of the mechanisms, which can be determined as:

$$X_{PZT}^{tr} = X_{PZT}^{nl}\cdot\frac{k_{pzt}}{k_{in} + k_{pzt}} \tag{14}$$

where k_{pzt} is the stiffness of the PZT. The stroke reduction can be neglected when the input stiffness of the mechanism is much smaller than the stiffness of PZT.

Consider again the flexure hinge b as an example, as shown in Figure 3c. The flexure hinge can be treated as a cantilever beam under combined loads at the free end. The maximum stress within the flexure hinge is the superposition of the axial and bending stress, which can be written as:

$$\sigma_{23}^{max} = \max_{x_3\in[0,l_2]} \left(\sigma_M + \sigma_N\right) \tag{15}$$

where $\sigma_N = \frac{F_{3x}}{w \cdot t_2}$ and $\sigma_M = \frac{6 \cdot M_{23}}{w \cdot t_2^2}$ are the axial stress and maximum bending stress of a cross-section within flexure hinge *b* at the position of x_3 with respect to node 3, respectively. For a general compliant bridge mechanism, the bending moment varies along the flexure hinge because of the hinge orientation. The moment can be deduced as:

$$M_{23} = M_3^{max} + F_{3y}^{max} \cdot x_3, (x_3 \in [0, l_2])$$ (16)

where M_3^{max} and F_{3y}^{max} are the maximum bending moment and shear force obtained by Equations (3)–(6) under the maximum input force of Equation (12). Similarly, the maximum stress within the flexure hinge *d* can be obtained as σ_{45}^{max}. The maximum stress in the compliant bridge mechanism can be determined as:

$$\sigma^{max} = \max \left(\sigma_{23}^{max}, \sigma_{45}^{max} \right)$$ (17)

3. Optimization

Using the established equations, piezo-driven compliant bridge mechanisms can be optimized for maximum displacement under geometric, stress, and stiffness constraints. Herein, a compliant bridge mechanism is optimized for use in a multiple degree of freedom positioner. Eight geometric parameters were investigated as variables, as listed in Table 1. The width of the mechanism was fixed at $w = 10$ mm. Aluminum alloy 7075-T6 was selected as the material with modulus of elasticity $E = 72$ GPa, a Poisson's ratio of $\mu = 0.33$, and modulus of shear obtained by $G = \frac{E}{2(1+\mu)}$.

Table 1. Boundary of the geometric parameter for optimization of the piezo-driven compliant bridge mechanism and the global optimal result.

Parameters (mm, °)	l_2	ffi_2	t_2	l_3	ffi_3	l_4	ffi_4	t_4
Upper boundary	20	45	2	20	45	20	45	2
Lower boundary	0.5	−45	0.4	0.5	−45	0.5	−45	0.4
Optimal result	1.96	4.07	0.4	8.1	4.01	1.96	4.07	0.4

During the optimization, the contours of the mechanism were constrained by:

$$\begin{cases} 0.0075 \text{ m} \leq l_2 \cos \delta_2 + l_3 \cos \delta_3 + l_4 \cos \delta_4 \leq 0.012 \text{ m} \\ -0.01 \text{ m} \leq l_2 \sin \delta_2 + l_3 \sin \delta_3 + l_4 \sin \delta_4 \leq 0.01 \text{ m} \end{cases}$$ (18)

The maximum stress is limited by:

$$\sigma_{max} \leq \frac{\sigma_u}{3}$$ (19)

where $\sigma_u = 505$ MPa is the ultimate strength of the material. In addition, a nominal actuation of 17.4 µm of the PZT and a preload of 40 N were employed. The input stiffness and output stiffness were constrained as:

$$\begin{cases} K_{in} \leq 7 \times 10^6 \text{ N/m} \\ K_{out} \geq 3.8 \times 10^4 \text{ N/m} \end{cases}$$ (20)

The objective function is specified by:

$$\text{Find max} : |da|$$ (21)

It can be predicted from Equation (9) that the optimization problem may have many local optima due to the underlying nonlinearity of the model. Therefore, instead of deriving a specific optimization method, a vast quantity of optimizations was carried out using the constrained nonlinear multivariable optimization function "fmincon" in MATLAB (R2013a, MathWorks, Natick, MA, USA) in this study. In each instance, the objective function, boundaries and constraints were the same as stated previously, whilst a random initial estimate within the parameter ranges was used.

3.1. Optimization Results

As shown in Figure 4, after using 300 solving instances with random initial estimates, the global maximum displacement amplification obtained by the optimization was around 12.8. In addition, various local optima were obtained which are greatly influenced by the initial estimates. The distributions of all the optima can be divided into four zones, as shown in Figure 4, where the quantity of instance from top to down are 70, 33, 181 and 16. The configuration of each instance is illustrated by plotting the central axis of the two flexure hinges and the middle link, as shown in Figure 5, where the origin of the coordinate system is set at node 2, with the x axis reverse to the input direction and y axis along the output direction. As can be seen, most samples in zone 1 are in aligned configurations, whilst most samples in zone 3 are in rhombic configurations. The optimal design in terms of displacement amplification under the constraints in this study is in the aligned configuration, and the optimal geometric parameters are determined as shown in Table 1.

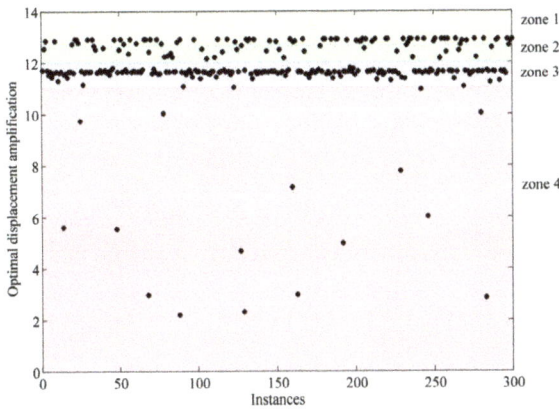

Figure 4. Distribution of the optimal displacement amplifications of the 300 instances with random initial estimations.

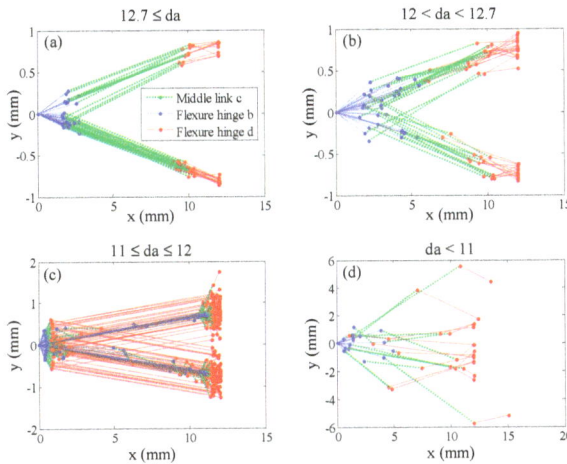

Figure 5. Illustrations of the configurations, zoned according to the values of the optimal displacement amplifications of the 300 instances with random initial estimations: (**a**) $12.7 \leq da$; (**b**) $12 < da < 12.7$; (**c**) $11 \leq da \leq 12$; (**d**) $da < 11$.

4. FEA and Experimental Evaluations

4.1. FEA

To verify the models and optimization, the mechanism obtained in previous section was further investigated using FEA and experiment. The model of the whole compliant bridge mechanism was constructed and analyzed within the ANSYS software package (15.0.7, ANSYS, Canonsburg, PA, USA). As shown in Figure 6a, a mesh model with 83,667 nodes and 42,955 elements was built, with refined mesh on the flexure parts. During the analyses, the bottom face of the mechanism was fixed, and a translational input force of 10 N is applied to the two input faces. The average displacements of the input and output faces were recorded, as shown in Table 2. The displacement amplification and input stiffness were then obtained. According to the input stiffness calculated by FEA, an input force of 152 N was actuated on the input faces to simulate the maximum PZT actuation of 17.4 μm with the preload of 40 N. The stress in such a situation was recorded as shown in Figure 6b. Then, in order to investigate the output stiffness, the output face was actuated by 10 N, while the input faces remained free. The results indicate that the deviations between the FEA and the analytical results are less than 11%.

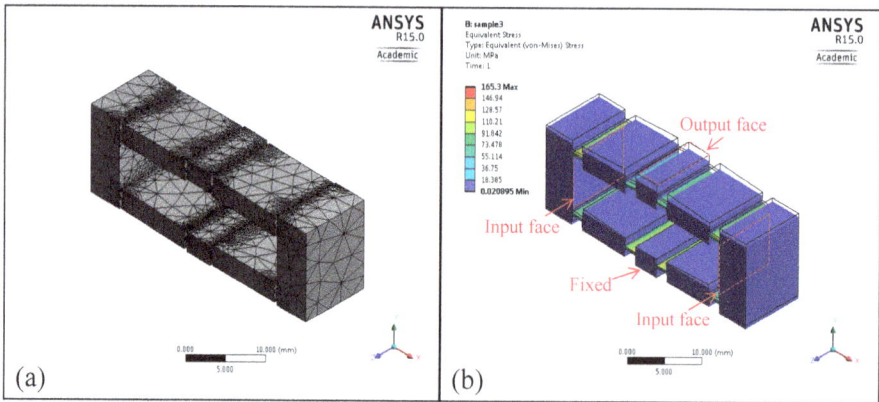

Figure 6. Finite element analysis of the global optimal compliant bridge mechanism: (**a**) mesh model; (**b**) maximum stress simulation.

Table 2. Performance of the global optimal compliant bridge mechanism by finite element analysis (FEA) and analytical equations.

Result	Y_{out} (μm)	X_{in} (μm)	K_{in} (N/m)	σ^{max} (MPa)	K_{out} (N/m)
FEA	18.6	1.56	6.43×10^6	165	3.9×10^4
Analytical	18.3	1.42	7.02×10^6	148	3.8×10^4
Deviation	1.5%	8.5%	9.2%	10.3%	2.5%

4.2. Experimental Evaluation

A prototype of the optimal compliant bridge mechanism was fabricated and tested, as shown in Figure 7. The prototype was manufactured from a piece of aluminum alloy 7075-T6 by wire-electrical discharging machining. A PZT (AE0505D16F, NEC, Tokyo, Japan) was inserted into the bridge mechanism and actuated by a controller (MDT693B, Thorlabs, Newton, NJ, USA). During the tests, the PZT was physically preloaded by two identical wedges which are placed together between the actuator and the input link of the compliant bridge mechanism. The PZT was adjusted and fastened manually, where the actuator could efficiently drive the input links. To ensure a constant actuation force during the experiments, the input stoke and the output displacements were tested under the same setting of preload. In the test of the input stroke, as shown in Figure 7a, one of the input links was fixed on the vibration-isolated table while the displacement of the other input link was measured by a position measuring probe (32.10924, TESA, North Kingstown, RI, USA) and read out by an analogue display (TTA20, TESA).The maximum input displacement measured was 13.5 μm. Then, the output displacement of the mechanism was tested as shown in Figure 7b, where the bottom face was mounted and the output displacements were measured by a laser interferometer (7003A, ZYGO, Berwyn, PA, USA). As shown in Figure 7c, the output displacement under the sinusoidal actuation was recorded and the detected maximum output displacement is 168 μm. As shown in Table 3, the analytical displacement amplification for the developed compliant bridge mechanism deviates less than 4% from the experimental result, and 8% with respect to the FEA result.

Table 3. Analytical, FEA and experimental results of displacement amplification for the developed compliant bridge mechanism.

Types of Result	FEA	Experimental	Analytical
da	11.95	12.44	12.86

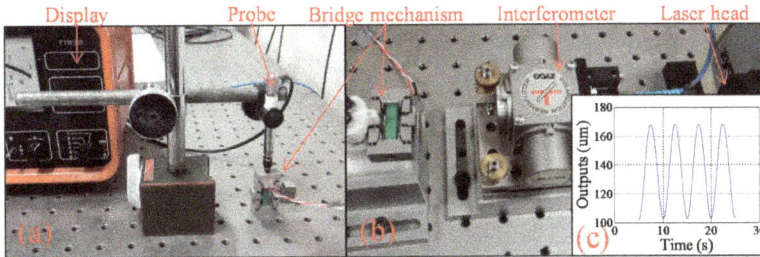

Figure 7. Photos of experimental apparatus: (**a**) setup of input stroke test; (**b**) setup of output displacement test; (**c**) outputs of sinusoidal actuations.

5. Comparisons with Previous Models

As shown in Figure 8, a general compliant bridge mechanism can be transformed into parallel, rhombic or aligned-type configurations by varying the six configuration parameters. By substituting the geometric characteristics of each configuration into the analytical equations, comparisons with previously developed models from the literature were carried out to investigate the feasibility of the models.

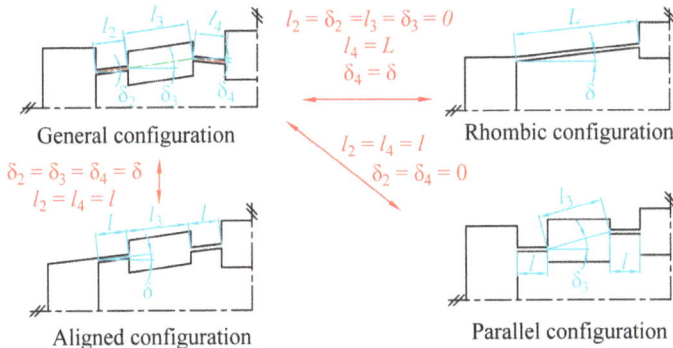

Figure 8. Variations of compliant bridge mechanisms between general, rhombic, parallel and aligned type configurations.

First, a parallel configuration can be represented within the general framework by:

$$\begin{cases} l_2 = l_4 = l \\ \delta_2 = \delta_4 = 0 \end{cases} \tag{22}$$

By substituting these configuration parameters into Equation (9), the general equation for displacement amplification can be written as:

$$da_{parallel} = \frac{\sin \delta_3 \cdot \left(c_{33}^l \cdot \cos \delta_3 \cdot l_3^2 + 2 \cdot c_{23}^l \cdot l_3 \right)}{c_{33}^l \cdot l_3^2 \cdot \cos^2 \delta_3 + 4 \cdot c_{11}^l} \tag{23}$$

where c_{ij}^l is the compliance factor of the flexure hinge corresponding to length l. Equation (23) is the same as that presented by Qi or Ling [23,26]. Secondly, the configuration parameters of the rhombic type compliant bridge mechanisms can be given as:

$$\begin{cases} l_2 = l_3 = \delta_2 = \delta_3 = 0 \\ l_4 = L \\ \delta_4 = \delta \end{cases} \tag{24}$$

By substituting the configuration parameters into Equation (9), the general equation of displacement amplification turns into:

$$da_{rhombic} = \frac{\sin(2 \cdot \delta) \cdot \left(2 \cdot c_{11}^L - \frac{c_{22}^L}{2} \right)}{c_{22}^L + 4 \cdot c_{11}^L \cdot \cos^2 \delta - c_{22}^L \cdot \cos^2 \delta} \tag{25}$$

where c_{ij}^L is the compliance factor of the flexure hinge corresponding to length L. Equation (25) is the same as that presented by Ling [26] (note that $c_{33}^L = \frac{3 \cdot c_{22}^L}{L^2}$ and $c_{23}^L = c_{32}^L = \frac{6 \cdot c_{22}^L}{4 \cdot L}$ have been applied as indicated in Equation (2) for strip type flexure hinges). Hence, it can be concluded that the presented models generalize both the parallel and rhombic type compliant bridge mechanism models that have been verified by previous studies. However, the equation for displacement amplification of the aligned-type compliant bridge mechanisms has not yet been investigated. The configuration parameters of the aligned-type compliant bridge mechanisms can be described as:

$$\begin{cases} \delta_2 = \delta_3 = \delta_4 = \delta \\ l_2 = l_4 = l \end{cases} \tag{26}$$

By substituting the configuration parameters into Equation (9), the equation for displacement amplification of aligned-type mechanisms is determined to be:

$$da_{aligned} = \frac{\sin \delta \cdot \cos \delta \cdot \left(4 \cdot c_{22}^l - 4 \cdot c_{11}^l + 2 \cdot c_{23}^l \cdot l_3 + 2 \cdot c_{32}^l \cdot l_3 + c_{33}^l \cdot l_3^2 \right)}{4 \cdot c_{22}^l \cdot \sin^2 \delta + 4 \cdot c_{11}^l \cdot \cos^2 \delta + 2 \cdot l_3 \cdot c_{23}^l \cdot \sin^2 \delta + 2 \cdot l_3 \cdot c_{32}^l \cdot \sin^2 \delta + l_3^2 \cdot c_{33}^l \cdot \sin^2 \delta} \tag{27}$$

Furthermore, numerical simulations were carried out to compare the presented equations with those proposed by Lobontiu [18] in terms of the six configuration parameters for general complaint bridge mechanisms. During the computations, only one parameter is varied in each analysis, while the other parameters were kept constant, as: $l_2 = l_4 = 0.002$ m, $l_3 = 0.02$ m, $\delta_2 = \delta_3 = \delta_4 = 5°$. The thickness and width of the flexure hinge are fixed at: $t_2 = t_4 = 0.0004$ m, $w = 0.004$ m. As shown in Figure 9, the results calculated by the proposed equations match well with those obtained by Lobontiu's equations This suggests that the presented models are feasible for compliant bridge mechanisms in general configurations for both macro and micro applications.

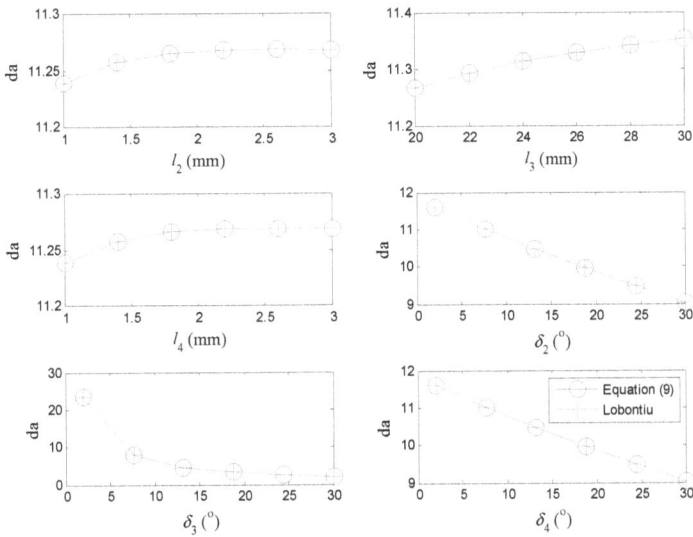

Figure 9. Numerical comparisons of displacement amplification between the new and Lobontiu's equations in terms of the configuration parameters.

6. Conclusions

In this study, a simplified analytical model for general compliant bridge mechanisms has been formulated based on beam theory and kinematic analysis. The model has been shown to accurately characterize compliant bridge mechanisms in parallel, aligned and rhombic type configurations. Analytical equations of input, output, displacement amplification, stiffness and stress have been obtained. The optimization of a piezo-driven compliant bridge mechanism has been accomplished based on the proposed models and equations. With the presented equations, optimizations can be achieved efficiently. The aligned configuration was found to be globally optimal within this framework. The optimal design was developed and investigated by FEA and experiment. The deviations between analytical displacement amplification and FEA and experiment are less than 8% and 4%, respectively. Comparisons with previous equations have indicated that the presented models are feasible for general compliant bridge mechanisms for both macro and micro applications. The equation for displacement amplification for aligned-type compliant bridge mechanisms was first obtained. The concise form of the proposed equations can help to facilitate the optimal design of compliant bridge mechanisms. Future work will be directed toward the nonlinear modeling of large deformation or material nonlinearity, dynamic modeling and precision control of the compliant bridge mechanisms.

Acknowledgments: This work was partially supported by the National Natural Science Foundation of China (Grant U1610111), PAPD project of Jiangsu Higher Education Institutions, Jiangsu Provincial Department of Education (Grant KYLX15_1421), Australian Research Council (ARC) LIEF (Grants LE0347024 and LE0775692), and ARC Discovery Projects (Grant DP140104019).

Author Contributions: Huaxian Wei conceived of the study, its coordination, and conducted the experiments. Wei Li and Bijan Shirinzadeh are the main supervisors, and Yuqiao Wang is the co-supervisor of this study. They provided the general direction, supervision, feedback and support of this research. Leon Clark provided the fundamentals of the experiments. Joshua Pinskier contributed to the analysis of experiment results and the framework of this study. All the authors helped in reviewing the manuscript until its final version.

Conflicts of Interest: The authors declare no conflict of interest.

Appendix A

The coefficients in Equations (8)−(11) are given as:

$$
\begin{aligned}
a_{11} = \; & \sin\delta_2\left(c_{22}^b\cdot\sin\delta_2 + \frac{c_{23}^b\cdot\left(c_{32}^d\cdot\sin\delta_4 - c_{32}^b\cdot\sin\delta_2 + c_{33}^d\cdot l_3\cdot\sin\delta_3\right)}{c_{33}^b + c_{33}^d}\right) \\
& + \sin\delta_4\left(c_{22}^d\cdot\sin\delta_4 + \frac{c_{23}^d\cdot\left(c_{32}^b\cdot\sin\delta_2 - c_{32}^d\cdot\sin\delta_4 + c_{33}^b\cdot l_3\cdot\sin\delta_3\right)}{c_{33}^b + c_{33}^d}\right) \\
& + c_{11}^b\cdot\cos^2\delta_2 + c_{11}^d\cdot\cos^2\delta_4 \\
& + \frac{l_3\cdot\sin\delta_3\cdot\left(c_{32}^b\cdot c_{33}^d\cdot\sin\delta_2 + c_{32}^d\cdot c_{33}^b\cdot\sin\delta_4 + c_{33}^b\cdot c_{33}^d\cdot l_3\cdot\sin\delta_3\right)}{c_{33}^b + c_{33}^d}
\end{aligned}
\tag{A1}
$$

$$
\begin{aligned}
a_{12} = \; & c_{11}^b\cdot\cos\delta_2\cdot\sin\delta_2 \\
& - \sin\delta_4\left(c_{22}^d\cdot\sin\delta_4 + \frac{c_{23}^d\cdot\left(c_{32}^b\cdot\cos\delta_2 - c_{32}^d\cdot\cos\delta_4 + c_{33}^b\cdot l_3\cdot\cos\delta_3\right)}{c_{33}^b + c_{33}^d}\right) \\
& - \sin\delta_2\left(c_{22}^b\cdot\sin\delta_2 + \frac{c_{23}^b\cdot\left(c_{32}^d\cdot\cos\delta_4 - c_{32}^b\cdot\cos\delta_2 + c_{33}^d\cdot l_3\cdot\cos\delta_3\right)}{c_{33}^b + c_{33}^d}\right) \\
& + c_{11}^d\cdot\cos\delta_4\cdot\sin\delta_4 \\
& - \frac{l_3\cdot\sin\delta_3\cdot\left(c_{32}^b\cdot c_{33}^d\cdot\cos\delta_2 + c_{32}^d\cdot c_{33}^b\cdot\cos\delta_4 + c_{33}^b\cdot c_{33}^d\cdot l_3\cdot\cos\delta_3\right)}{c_{33}^b + c_{33}^d}
\end{aligned}
\tag{A2}
$$

$$
\begin{aligned}
a_{21} = \; & \cos\delta_2\left(c_{22}^b\cdot\sin\delta_2 + \frac{c_{23}^b\cdot\left(c_{32}^d\cdot\sin\delta_4 - c_{32}^b\cdot\sin\delta_2 + c_{33}^d\cdot l_3\cdot\sin\delta_3\right)}{c_{33}^b + c_{33}^d}\right) \\
& + \cos\delta_4\left(c_{22}^d\cdot\sin\delta_4 + \frac{c_{23}^d\cdot\left(c_{32}^b\cdot\sin\delta_2 - c_{32}^d\cdot\sin\delta_4 + c_{33}^b\cdot l_3\cdot\sin\delta_3\right)}{c_{33}^b + c_{33}^d}\right) \\
& - c_{11}^b\cdot\cos\delta_2\cdot\sin\delta_2 - c_{11}^d\cdot\cos\delta_4\cdot\sin\delta_4 \\
& + \frac{l_3\cdot\sin\delta_3\cdot\left(c_{32}^b\cdot c_{33}^d\cdot\sin\delta_2 + c_{32}^d\cdot c_{33}^b\cdot\sin\delta_4 + c_{33}^b\cdot c_{33}^d\cdot l_3\cdot\sin\delta_3\right)}{c_{33}^b + c_{33}^d}
\end{aligned}
\tag{A3}
$$

$$
\begin{aligned}
a_{22} = \; & -\frac{l_3\cdot\cos\delta_3\cdot\left(c_{32}^b\cdot c_{33}^d\cdot\cos\delta_2 + c_{32}^d\cdot c_{33}^b\cdot\cos\delta_4 + c_{33}^b\cdot c_{33}^d\cdot l_3\cdot\cos\delta_3\right)}{c_{33}^b + c_{33}^d} - c_{11}^b\cdot\sin^2\delta_2 \\
& - c_{11}^d\cdot\sin^2\delta_4 \\
& - \cos\delta_2\left(c_{22}^b\cdot\cos\delta_2 + \frac{c_{23}^b\cdot\left(c_{32}^d\cdot\cos\delta_4 - c_{32}^b\cdot\cos\delta_2 + c_{33}^d\cdot l_3\cdot\cos\delta_3\right)}{c_{33}^b + c_{33}^d}\right) \\
& - \cos\delta_4\left(c_{22}^d\cdot\cos\delta_4 + \frac{c_{23}^d\cdot\left(c_{32}^b\cdot\cos\delta_2 - c_{32}^d\cdot\cos\delta_4 + c_{33}^b\cdot l_3\cdot\cos\delta_3\right)}{c_{33}^b + c_{33}^d}\right)
\end{aligned}
\tag{A4}
$$

References

1. Tian, Y.; Shirinzadeh, B.; Zhang, D.; Liu, X.; Chetwynd, D.G. Design and forward kinematics of the compliant micro-manipulator with lever mechanisms. *Precis. Eng.* **2009**, *33*, 466–475. [CrossRef]
2. Chen, T.; Wang, Y.; Yang, Z.; Liu, H; Liu, J.; Sun, L. A PZT Actuated Triple-Finger Gripper for Multi-Target Micromanipulation. *Micromachines* **2017**, *8*, 33. [CrossRef]
3. Bhagat, U.; Shirinzadeh, B.; Clark, L.; Chea, P.; Qin, Y.; Tian, Y.; Zhang, D. Design and analysis of a novel flexure-based 3-DOF mechanism. *Mech. Mach. Theory* **2014**, *74*, 173–187. [CrossRef]
4. Zhou, M.; Fan, Z.; Ma, Z.; Zhao, H.; Guo, Y.; Hong, K.; Li, Y.; Liu, H.; Wu, D. Design and Experimental Research of a Novel Stick-Slip Type Piezoelectric Actuator. *Micromachines* **2017**, *8*, 150. [CrossRef]
5. Chen, X.; Li, Y. Design and analysis of a new high precision decoupled XY compact parallel micromanipulator. *Micromachines* **2017**, *8*, 1–13. [CrossRef]
6. Qin, Y.; Shirinzadeh, B.; Zhang, D.; Tian, Y. Design and Kinematics Modeling of a Novel 3-DOF Monolithic Manipulator Featuring Improved Scott-Russell Mechanisms. *J. Mech. Des.* **2013**, *135*. [CrossRef]
7. Wei, H.; Li, W.; Liu, Y.; Wang, Y.; Yang, X. Quasi-static analysis of a compliant tripod stage with plane compliant lever mechanism. *Proc. Inst. Mech. Eng. Part C J. Mech. Eng. Sci.* **2017**, *231*, 1639–1650. [CrossRef]
8. Zubir, M.N.M.; Shirinzadeh, B.; Tian, Y. A new design of piezoelectric driven compliant-based microgripper for micromanipulation. *Mech. Mach. Theory* **2009**, *44*, 2248–2264. [CrossRef]

9.   Zhao, Y.; Zhang, C.; Zhang, D.; Shi, Z.; Zhao, T. Mathematical Model and Calibration Experiment of a Large Measurement Range Flexible Joints 6-UPUR Six-Axis Force Sensor. *Sensors* **2016**, *16*, 1271. [CrossRef] [PubMed]
10.   Howell, L.L.; Midha, A. Parametric Deflection Approximations for End-Loaded, Large-Deflection Beams in Compliant Mechanisms. *J. Mech. Des.* **1995**, *117*, 156–165. [CrossRef]
11.   Tian, Y.; Shirinzadeh, B.; Zhang, D. Closed-form compliance equations of filleted V-shaped flexure hinges for compliant mechanism design. *Precis. Eng. J. Int. Soc. Precis. Eng. Nanotechnol.* **2010**, *34*, 408–418. [CrossRef]
12.   Qin, Y.; Shirinzadeh, B.; Tian, Y.; Zhang, D.; Bhagat, U. Design and computational optimization of a decoupled 2-DOF monolithic mechanism. *IEEE/ASME Trans. Mechatron.* **2014**, *19*, 872–881. [CrossRef]
13.   Li, Y.; Xu, Q. Design and robust repetitive control of a new parallel-kinematic XY piezostage for micro/nanomanipulation. *IEEE/ASME Trans. Mechatron.* **2012**, *17*, 1120–1132. [CrossRef]
14.   Clark, L.; Shirinzadeh, B.; Zhong, Y.; Tian, Y.; Zhang, D. Design and analysis of a compact flexure-based precision pure rotation stage without actuator redundancy. *Mech. Mach. Theory* **2016**, *105*, 129–144. [CrossRef]
15.   Pinskier, J.; Shirinzadeh, B.; Clark, L.; Qin, Y.; Fatikow, S. Design, development and analysis of a haptic-enabled modular flexure-based manipulator. *Mechatronics* **2016**, *40*, 156–166. [CrossRef]
16.   Pokines, B.J.; Garcia, E. A smart material microamplification mechanism fabricated using LIGA. *Smart Mater. Struct.* **1999**, *7*, 105–112. [CrossRef]
17.   Ma, H.; Yao, S.; Wang, L.; Zhong, Z. Analysis of the displacement amplification ratio of bridge-type flexure hinge. *Sensors Actuators A Phys.* **2006**, *132*, 730–736. [CrossRef]
18.   Lobontiu, N.; Garcia, E. Analytical model of displacement amplification and stiffness optimization for a class of flexure-based compliant mechanisms. *Comput. Struct.* **2003**, *81*, 2797–2810. [CrossRef]
19.   Kim, J.H.; Kim, S.H.; Kwak, Y.K. Development of a piezoelectric actuator using a three-dimensional bridge-type hinge mechanism. *Rev. Sci. Instrum.* **2003**, *74*, 2918–2924. [CrossRef]
20.   Ye, G.; Li, W.; Wang, Y.; Yang, X.; Yu, L. Kinematics analysis of bridge-type micro-displacement mechanism based on flexure hinge. *IEEE Int. Conf. Inf. Autom.* **2010**, 66–70. [CrossRef]
21.   Borboni, A.; Faglia, R. Stochastic Evaluation and Analysis of Free Vibrations in Simply Supported Piezoelectric Bimorphs. *J. Appl. Mech.* **2013**, *80*, 21003. [CrossRef]
22.   Borboni, A.; De Santis, D. Large deflection of a non-linear, elastic, asymmetric Ludwick cantilever beam subjected to horizontal force, vertical force and bending torque at the free end. *Meccanica* **2014**, *49*, 1327–1336. [CrossRef]
23.   Qi, K.; Xiang, Y.; Fang, C.; Zhang, Y.; Yu, C. Analysis of the displacement amplification ratio of bridge-type mechanism. *Mech. Mach. Theory* **2015**, *87*, 45–56. [CrossRef]
24.   Mottard, P.; St-Amant, Y. Analysis of flexural hinge orientation for amplified piezo-driven actuators. *Smart Mater. Struct.* **2009**, *18*, 35005. [CrossRef]
25.   Shao, S.; Xu, M.; Zhang, S.; Xie, S. Stroke maximizing and high efficient hysteresis hybrid modeling for a rhombic piezoelectric actuator. *Mech. Syst. Signal Process.* **2016**, *75*, 631–647. [CrossRef]
26.   Ling, M.; Cao, J.; Zeng, M.; Lin, J.; Inman, D.J. Enhanced mathematical modeling of the displacement amplification ratio for piezoelectric compliant mechanisms. *Smart Mater. Struct.* **2016**, *25*, 75022. [CrossRef]
27.   Kim, J.H.; Kim, S.H.; Kwak, Y.K. Development and optimization of 3-D bridge-type hinge mechanisms. *Sensors Actuators A Phys.* **2004**, *116*, 530–538. [CrossRef]
28.   Ni, Y.; Deng, Z.; Li, J.; Wu, X.; Li, L. Multi-Objective Design Optimization of an Over-Constrained Flexure-Based Amplifier. *Algorithms* **2015**, *8*, 424–434. [CrossRef]
29.   Jung, H.J.; Kim, J.H. Novel piezo driven motion amplified stage. *Int. J. Precis. Eng. Manuf.* **2014**, *15*, 2141–2147. [CrossRef]
30.   Lobontiu, N. *Compliant Mechanisms: Design of Flexure Hinges*; CRC Press: Boca Raton, FL, USA, 2002; ISBN 1420040278.

micromachines

MDPI

Article

Influences of Excitation on Dynamic Characteristics of Piezoelectric Micro-Jets

Kai Li [1], Jun-Kao Liu [1,*], Wei-Shan Chen [1] and Lu Zhang [2]

[1]   State Key Laboratory of Robotics and System, Harbin Institute of Technology, Harbin 150001, China;
      sdcxlikai@126.com (K.L.); cws@hit.edu.cn (W.-S.C.)
[2]   Aero Engine Corporation of China, Harbin Dongan Engine Corporation LTD, Harbin 150001, China;
      zhanglu916hit@163.com
*   Correspondence: jkliu@hit.edu.cn; Tel.: +86-451-8641-6119

Received: 8 May 2017; Accepted: 30 June 2017; Published: 5 July 2017

Abstract: Piezoelectric micro-jets are based on piezoelectric ink-jet technology and can achieve the drop-on demand requirements. A piezoelectric micro-jet which is designed for bearing lubrication is presented in this paper. In order to analyze the fluid dynamic characteristics of the piezoelectric micro-jet so as to obtain good injection performance, a direct coupling simulation method is proposed in this paper. The effects of inlet and viscous losses in the cavity are taken into account, which are close to the actual conditions in the direct coupling method. The effects of the pulse excitation parameters on the pinch-off time, tail length, velocity, and volume of the droplet are analyzed by the proposed direct coupling method. The pressure distribution inside the cavity of the micro-jet and the status of the droplet formation at different times are also given. In addition, the method is proved to be effective in predicting and analyzing the fluid dynamic characteristics of piezoelectric micro-jets by comparing the simulation results with the experimental results.

Keywords: coupling analysis; fluid dynamic; piezoelectric micro-jet

1. Introduction

As it can achieve drop-on-demand injection, piezoelectric micro-jet technology is widely applied in various fields, such as ejecting metal nanoparticles [1], cell printing [2–4], color printing [5], drug delivery [6–8], manufacturing [9–13], and sensors [14–17], etc. In addition, there are other alternate technologies to achieve drop-on-demand injection such as pyroelectric ink-jet printing [18–22], etc. In order to improve the injection performance, the influences of excitation parameters on fluid dynamic characteristics should be analyzed. Researchers have studied these dynamic characteristics by experiments and obtained the actual injection performance of the micro-jet [23–31]. Also, simulations have been used to study the characteristics of the micro-jet, as they can help to understand the outcomes of experiments and can help to pave the way to new designs. Of course, such designs should and must be checked experimentally. Many simulation projects have been carried out, but piezoelectric micro-jets with small cavities are often neglected in simulations, and only the dynamic characteristics of the nozzle part are analyzed [32–38]. For piezoelectric micro-jets with large cavities, such as the lubricating micro-jet presented in this paper, the influences of oil inlet and viscosity loss in the cavity cannot be neglected. In our previous study, an indirect simulation method was proposed to analyze the dynamic characteristics of the nozzle part of the piezoelectric micro-jet for lubricating, and although the results were consistent with the experimental results, the accuracy was not ideal when the quantitative analyses were carried out [39]. Therefore, a direct coupling simulation method, which takes the viscous loss and inlet effect into account, is proposed in this paper. Then, quantitatively analyses are carried out to study the influences of pulse voltage parameters on the injection performance, and the validity of this method is proved by the experiments. As some air will be trapped as bubbles in the cavity, which reduces the injection performance of the piezoelectric micro-jet [40], here the influences of bubbles are not discussed.

As the designed micro-jet is used for bearing lubrication, for a single time oil supply, the required oil volume should be met by controlling the ejected number of oil droplets. Moreover, the response time of the lubrication, which affects the timeliness of lubrication, is effected by the velocity of droplets. Thus, the volume and velocity of oil droplets are key parameters. In this paper, we obtained the volumes of the droplets under a single voltage pulse; then, we were able to determine the required number of voltage pulses for different oil volume requirements. In general, the piezoelectric micro-jet is designed to be embedded in the bearing system, and the nozzles are located between the inner and outer raceways of the bearing system. All of the ejected oil droplets will eventually be transported to the raceways with the assistance of the balls; thus, the directionality and precision of landing place are not discussed in this paper. The volume of the droplets is discussed here, as are the total volume including the volumes of satellite droplets. In order to improve the uniformity of lubrication, two nozzles are symmetrical designed.

2. Working Principle

The piezoelectric micro-jet analyzed here is designed for the lubrication of a bearing system; the micro-jets are embedded in the bearing system without increasing the mass and volume of the whole system, as described in our previous work [39]. When high levels of the pulse voltages are applied to the piezoelectric vibrator, the droplets are ejected out by the positive pressure waves created in the cavity, and some air enters into the cavity from the nozzle. Then, low levels of the pulse voltages are applied, the vibrator is restored to its original state, negative pressure is created in the cavity, and the cavity is refilled by the oil from the inlet, while some of the air is squeezed out as well. Thus, we maintain the back-pressure a little higher than the atmospheric pressure by a dispenser, and the lubricating oil will not be pushed out due to its high viscosity and the small nozzle size, the back-pressure is verified in experiments (about 500 to 1000 Pa). In the simulations, the initial pressure boundary condition of the inlet is set as zero and a constant value (500 Pa) is set after one pulse excitation to simulation the function of the dispenser, while the volume fraction of the inlet is set as 1, namely, the inlet part is filled with oil.

According to the indirect coupling method proposed in our previous work, the displacements of particles on the vibrator at different positions and timepoints should be obtained first, following which the inlet velocities of the nozzle are calculated, which depend on the derived model by calculating the volume changes in the cavity. The indirect coupling method assumes that the fluid is completely incompressible, while the impact of the lubricating supply is ignored. For the direct coupling method proposed here, the velocities of particles at different positions and timepoints are applied as the velocity boundary conditions of the fluid-solid coupling interface. Moreover, the influence of the lubricating supply is taken into account, and the method can be used to simulate a compressible fluid. The model used in the simulations is shown in Figure 1; the coupling interface is between the copper substrate and the lubricating oil domain. The piezoelectric micro-jet is a central symmetrical rotating device; Figure 1 illustrates the sectional view of the device for two-dimensional simplification. In order to reduce the contact between the droplets and the outermost shell during the molding processes, the nozzle is designed as conical. Furthermore, because of the complex processing of machining the cone-shape in the shell, we selected inverse cone instead. The diameter of the nozzle is designed as 0.1 mm.

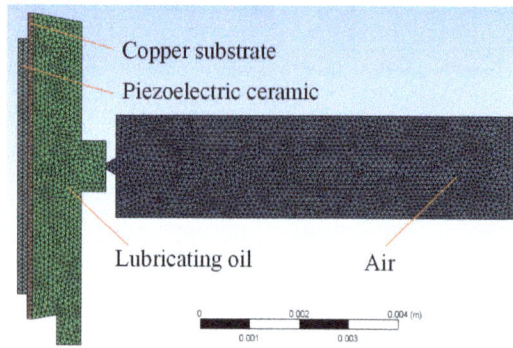

Figure 1. The model used in simulations.

As the piezoelectric vibrator is centrally symmetrical (as shown in our two-dimensional simplification of the design), here we use R to represent the radial distance of the particle on the vibrator to its symmetry axis. The resonance frequency of the vibrator is largely affected by the lubricating oil due to the coupling effect. Here, the resonance frequency of the vibrator is 1.83 kHz, which is obtained by the concerned coupling effect. Therefore, the driving frequency is set as 1.83 kHz as well. The velocities of the particle (with radial distance $R = 15$ mm) at different timepoints are shown in Figure 2. As it can be seen in the figure, the vibration tends to be stable after 15 cycles. Thus, in order to simulate the fluid dynamic characteristics of the stable work of the micro-jet, the particle velocities at the steady state are chosen as the velocity inlet boundary condition of the coupling interface.

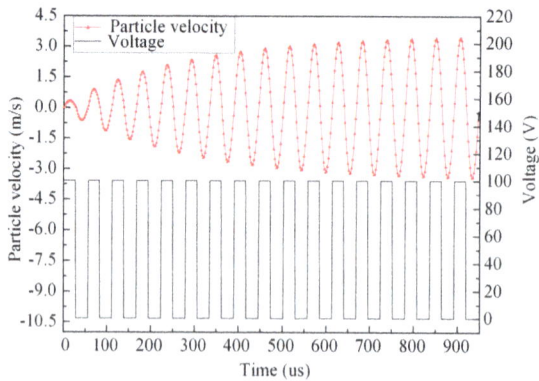

Figure 2. The particle velocity and its corresponding pulse voltage.

The relationship between interface particle velocities and time is shown in Figure 3. As shown in the figure, the maximum particle velocity occurs at the center (radial distance $R = 15$ mm) of the interface. Due to the effect of inertia force, the particle will have a reverse speed when the pulse excitation is removed. The velocity data at different positions and timepoints is used as the velocity inlet boundary conditions of the fluid-solid coupling interface in the next two-phase flow simulations.

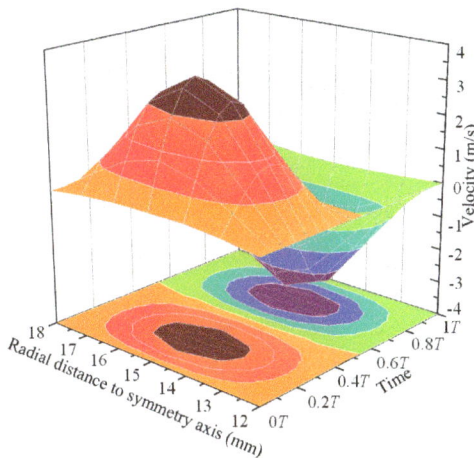

Figure 3. The velocities of the particles in the fluid-solid coupling interface.

The fluid dynamic characteristics of the micro-jet are analyzed by using the finite element software Fluent, and the two-dimensional model of the micro-jet is shown in Figure 1. The volume of fluid (VOF) model is selected to solve the two-phase flow problem. The density and viscosity of the lubricating oil are set as 859 kg/m^3 and 8.91 × 10^{-3} Pa·s, respectively. The density and viscosity of the air is set as 1.225 kg/m^3 and 1.7894 × 10^{-5} Pa·s, respectively. The air is set as the primary phase and the continuum surface force model is selected with a constant surface tension of 0.041 N/m. The boundary condition of the fluid-solid coupling interface is set as velocity-inlet, and the velocity values at different positions depend on the particle velocity data obtained from the vibration simulations of the vibrator. The boundary condition of the oil supply inlet is set as velocity-inlet with a constant value of 0. The contact angel of the inner wall of the cavity and the outer wall of the nozzle are set as 85° and 165°, respectively. The boundary condition of the outlet in air part is set as pressure-outlet with a constant pressure value of 0.

3. Analysis Results

In order to obtain the optimal pulse voltages to improve the injection performance, the influences of pulse excitation parameters on the fluid dynamic characteristics are analyzed. When the duty ratio and amplitude of the pulse voltages are set as 0.5 and 100 V, respectively, the pressure distributions in the cavity of the micro-jet during a period are shown in Figure 4, where T is the period. It can be see that, with the increase of time, the pressure of the nozzle increases before 0.25T and then decreases after it, which is consistent with the loaded pulse voltage. After 0.5T, the nozzle pressure value gradually changes to negative and the absolute value reaches the maximum at nearly 0.75T, which is due to the effect of the inertia force of the vibrator.

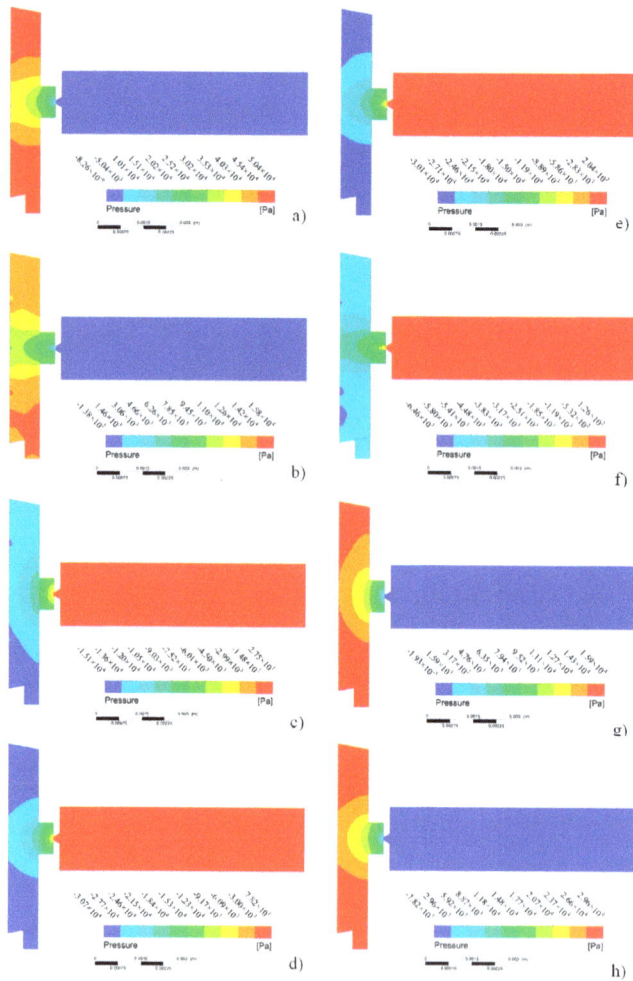

Figure 4. The pressure distribution at different timepoints: (**a**) 0.12*T*; (**b**) 0.25*T*; (**c**) 0.38*T*; (**d**) 0.5*T*; (**e**) 0.64*T*; (**f**) 0. 75*T*; (**g**) 0.88*T*; (**h**) 1*T*.

The parameters which are related to the injection performance of the micro-jet mainly include the pinch-off time (the time at which the droplet is completely ejected out of the nozzle), which reflects the response time; the tail length (the moving distance of the droplet over the pinch-off time), which determines the minimum working distance; the velocity of the droplet, which reflects the injection intensity; and the volume of the droplet, which reflects the injection precision. By the direct coupling method, the velocity and pinch-off time of the droplet can be obtained directly; however, the tail length and the volume of the droplet must be obtained through calculations.

As shown in Figure 5, the injection status of the micro-jet at the pinch-off time can be obtained directly, then the tail length can be calculated based on the scale and measurements. The red part represents the air and the blue part represents the oil. Furthermore, as the velocity of the droplet tail is large than zero, the tail of the droplet is ejected out as well. The tail of the droplet is formed as satellite droplets. Since the model in the simulation is a two-dimensional model, the three-dimensional volume of the droplet cannot be gained directly. We establish the three-dimensional model of the droplet in relative software based on its two-dimensional model. As the volume of the liquid droplets are calculated approximately by the results of the two-dimensional model, the calculated values are larger than the actual volumes. Thus, the volume results obtained from modeling software should be multiplied by 0.613~0.684 according to our experience; here, we take 0.63.

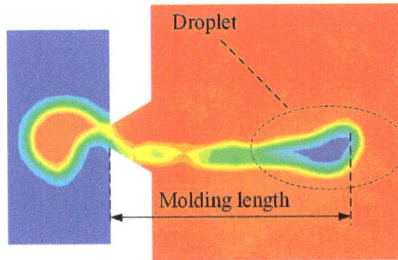

Figure 5. The model for the calculation of the volume and tail length of a droplet.

3.1. The Influences of the Voltage Amplitude

The pinch-off time and droplet velocities obtained by the direct coupling method varies with different amplitudes of pulse voltages, as shown in Figure 6. From this figure we can see that, with the increase of the voltage amplitude, the pinch-off time decreases gradually and the average change rate is $-0.2976\ \mu s \cdot V^{-1}$. The droplet velocity increases nearly lineally with the increase of the voltage amplitude, and the average change rate is $0.0403\ m \cdot s^{-1} \cdot V^{-1}$. Thus, with the increase of the voltage amplitude, both the injection intensity and the injection response speed of the micro-jet are enhanced.

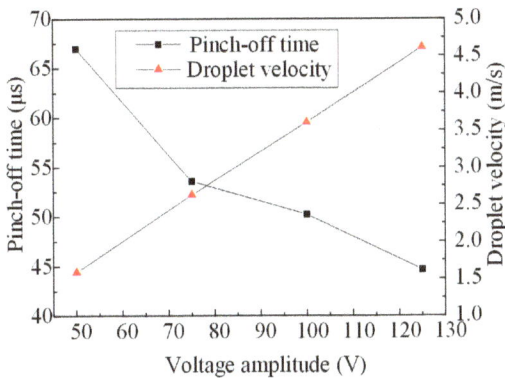

Figure 6. The pinch-off time and droplet velocity varies with different amplitudes of pulse voltages.

The tail length varies with different voltage amplitudes which, as obtained by the direct coupling method, are shown in Figure 7. From this figure, it can be seen that the tail length increases along with the increase of the voltage amplitude, and the average change rate is $0.0107\ mm \cdot V^{-1}$.

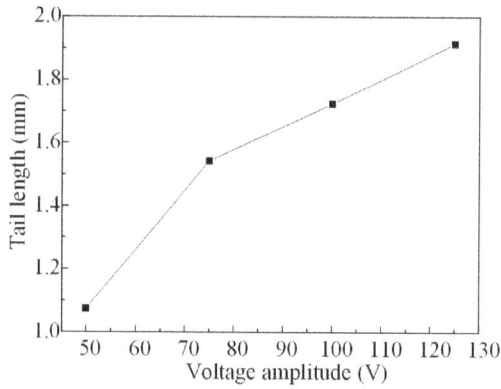

Figure 7. The tail length varies with different voltage amplitudes.

The injection status varies with different voltage amplitudes, which are obtained by the method of direct coupling, as shown in Figure 8. It can be seen that, when the voltage amplitude is relatively low, the ejected droplets are not symmetrical. The reason for this is that the effect of the oil supply inlet is taken into account in the direct coupling method, which results in an unsymmetrical pressure distribution at the nozzle. Furthermore, with the increase of the voltage amplitude, the effect is faded gradually.

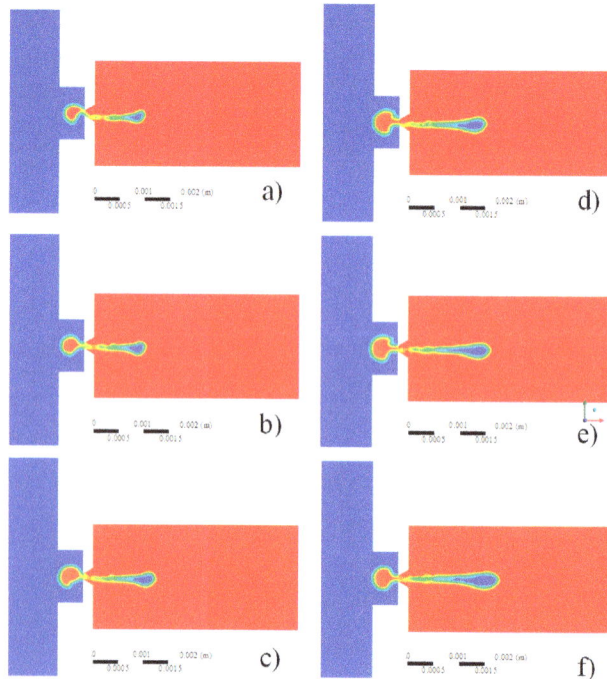

Figure 8. The injection status varies with different voltage amplitudes: (**a**) 50 V; (**b**) 60 V; (**c**) 75 V; (**d**) 90 V; (**e**) 100 V; (**f**) 125 V.

3.2. The Influence of the Duty Ratio

Another important parameter of the pulse voltage is the duty ratio; the pinch-off time and droplet velocity vary with different duty ratios obtained from the direct coupling method, as shown in Figure 9. From this figure we can see that, along with the increase of the duty ratio, the pinch-off time increases nearly linearly and the droplet velocity decreases. The change rate of the pinch-off time and droplet velocity are 37.2 µs and −5.291 m/s, respectively. Therefore, with the increase of the duty ratio, both the injection intensity and the injection response speed of the micro-jet are weakened.

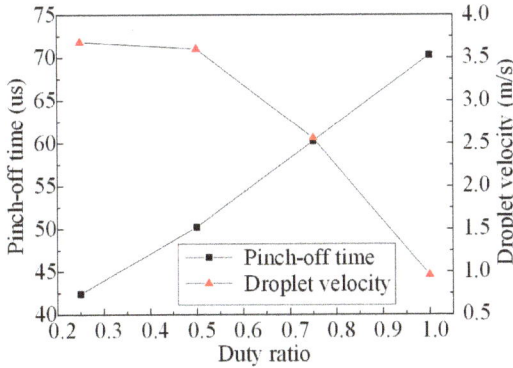

Figure 9. The pinch-off time and droplet velocity vary with different duty ratios.

With the increase of the duty ratio, the change curve of the tail length is obtained by the direct coupling method, as shown in Figure 10. From this figure we can see that, according to the results, the tail length increases before the turning point (duty ratio $\alpha = 0.6$) and then decreases. The reason for this is that, according to the directing coupling method, with the increase of the duty ratio, the energy dissipation increases after the turning point (duty ratio $\alpha = 0.6$) and the droplets are molded after a shorter moving distance as a result of the effect of the asymmetrical pressure at the nozzle.

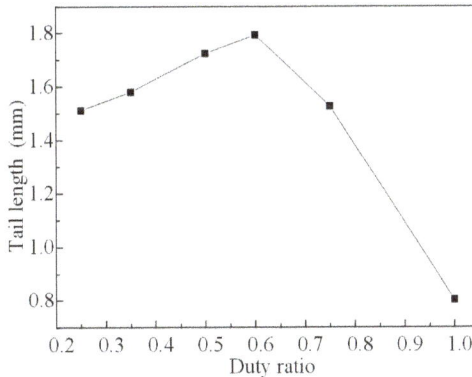

Figure 10. The tail length varies with different duty ratios.

As shown in Figure 11, with the increase of the duty ratio, the injection intensity increases before the turning point (duty ratio $\alpha = 0.6$) and then decreases after it. The reason for this is that the energy loss created in the cavity, which is related to the viscosity of the lubricating oil, increases with the increase of the duty ratio.

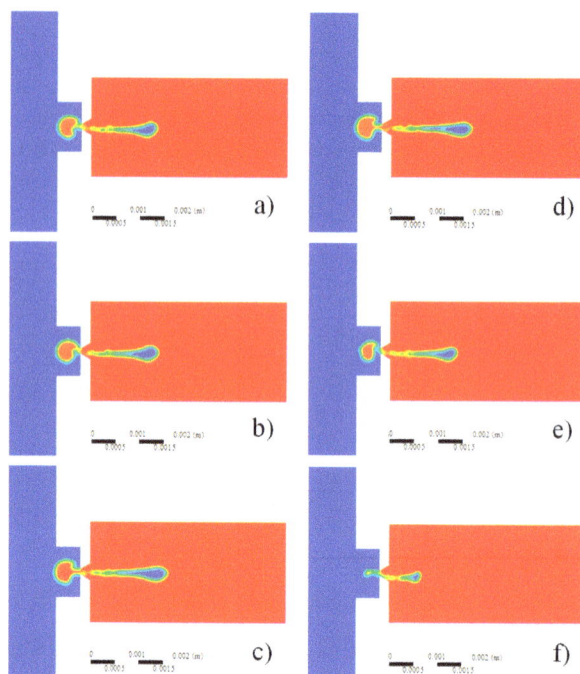

Figure 11. The injection status varies with different duty ratios: (**a**) α = 0.25; (**b**) α = 0.35; (**c**) α = 0.5; (**d**) α = 0.6; (**e**) α = 0.75; (**f**) α = 0.99.

3.3. Comparison with the Experiment Results

The experimental platform and method are described in our previous work [39]. The experimental platform mainly consists of a waveform generator, digital storage oscilloscope, and power amplifier. The droplet volume varies with different duty ratios, which are obtained by experiments and simulations. As shown in Figure 12, we can see that with the increase of the duty ratio, the droplet volume increases gradually (when the duty ratio α < 0.6). The results of the simulation are consonant with the experimental results, and the errors are acceptable for quantitative analyses.

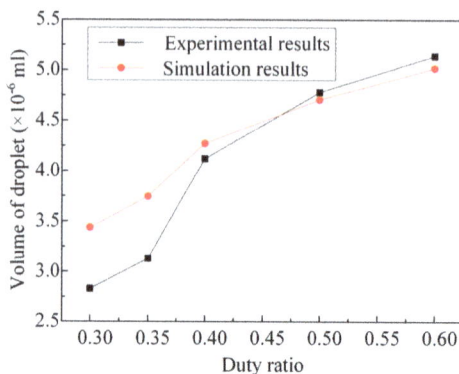

Figure 12. The droplet volume varies with different duty ratios.

4. Conclusions

A direct coupling method which is used to simulate the fluid dynamic characteristic of the piezoelectric micro-jet is proposed in this paper, and the model built according to this method is closer to the actual structure of the micro-jet. The effect of the oil supply inlet is taken into account in the direct coupling method, and it has been proven that this effect cannot be ignored when analyzing the influence of the voltage parameters on the injection performance. The change rate of the pinch-off time, droplet velocity, and tail length are given, which can be used as the references for adjusting the injection performance by modifying the parameters of pulse voltages. Comparing the results obtained from the direct coupling method with the experimental data, it is demonstrated that the direct coupling method proves to be an effective and feasible method to quantitatively simulate the fluid dynamic characteristics of the micro-jet.

In future research, we will carry out three-dimensional simulations, as the volume of droplets can be obtained directly from the results of a three-dimensional simulation analysis, and the model is closer to reality, namely, the simulations will be more accurate.

Acknowledgments: This project is supported by the National Natural Science Foundation of China (grant number 51075082, 51375107).

Author Contributions: Kai Li, Jun-Kao Liu and Wei-Shan Chen designed the experiments; Kai Li, Jun-Kao Liu and Lu Zhang performed the experiments and analyzed the data; Kai Li and Lu Zhang wrote the paper.

Conflicts of Interest: The authors declare no conflict of interest.

References

1. Torabi, P.; Petros, M.; Khoshnevis, B. Enhancing the resolution of selective inhibition sintering (sis) for metallic part fabrication. *Rapid Prototyp. J.* **2015**, *21*, 186–192. [CrossRef]
2. Cheng, E.; Yu, H.R.; Ahmadi, A.; Cheung, K.C. Investigation of the hydrodynamic response of cells in drop on demand piezoelectric inkjet nozzles. *Biofabrication* **2016**, *8*. [CrossRef] [PubMed]
3. Lorber, B.; Hsiao, W.K.; Martin, K.R. Three-dimensional printing of the retina. *Curr. Opin. Ophthalmol.* **2016**, *27*, 262–267. [CrossRef] [PubMed]
4. Tse, C.; Whiteley, R.; Yu, T.; Stringer, J.; MacNeil, S.; Haycock, J.W.; Smith, P.J. Inkjet printing schwann cells and neuronal analogue NG108–15 cells. *Biofabrication* **2016**, *8*. [CrossRef] [PubMed]
5. Lee, F.; Mills, R.; Talke, F. The application of drop-on-demand ink jet technology to color printing. *IBM J. Res. Dev.* **1984**, *28*, 307–313. [CrossRef]
6. Allen, E.A.; O'Mahony, C.; Cronin, M.; O'Mahony, T.; Moore, A.C.; Crean, A.M. Dissolvable microneedle fabrication using piezoelectric dispensing technology. *Int. J. Pharm.* **2016**, *500*, 1–10. [CrossRef] [PubMed]
7. Boehm, R.D.; Jaipan, P.; Skoog, S.A.; Stafslien, S.; VanderWal, L.; Narayan, R.J. Inkjet deposition of itraconazole onto poly(glycolic acid) microneedle arrays. *Biointerphases* **2016**, *11*. [CrossRef] [PubMed]
8. Scoutaris, N.; Chai, F.; Maurel, B.; Sobocinski, J.; Zhao, M.; Moffat, J.G.; Craig, D.Q.; Martel, B.; Blanchemain, N.; Douroumis, D. Development and biological evaluation of inkjet printed drug coatings on intravascular stent. *Mol. Pharm.* **2016**, *13*, 125–133. [CrossRef] [PubMed]
9. Chen, J.-J.; Lin, G.-Q.; Wang, Y.; Sowade, E.; Baumann, R.R.; Feng, Z.-S. Fabrication of conductive copper patterns using reactive inkjet printing followed by two-step electroless plating. *Appl. Surf. Sci.* **2017**, *396*, 202–207. [CrossRef]
10. Jiang, J.; Bao, B.; Li, M.; Sun, J.; Zhang, C.; Li, Y.; Li, F.; Yao, X.; Song, Y. Fabrication of transparent multilayer circuits by inkjet printing. *Adv. Mater.* **2016**, *28*, 1420–1426. [CrossRef] [PubMed]
11. Tomaszewski, G.; Potencki, J. Drops forming in inkjet printing of flexible electronic circuits. *Circ. World* **2017**, *43*, 13–18. [CrossRef]
12. Korvink, J.G.; Smith, P.J.; Shin, D.Y. *Inkjet-Based Micromanufacturing*; John Wiley & Sons, Inc.: Hoboken, NJ, USA, 2012.
13. Hutchings, I.M.; Martin, G.D. *Inkjet Technology for Digital Fabrication*; John Wiley & Sons, Inc.: Hoboken, NJ, USA, 2012.

14. Fuchiwaki, Y.; Yabe, Y.; Adachi, Y.; Tanaka, M.; Abe, K.; Kataoka, M.; Ooie, T. Inkjet monitoring technique with quartz crystal microbalance (QCM) sensor for highly reproducible antibody immobilization. *Sens. Actuators A* **2014**, *219*, 1–5. [CrossRef]

15. Sielmann, C.J.; Busch, J.R.; Stoeber, B.; Walus, K. Inkjet printed all-polymer flexural plate wave sensors. *IEEE Sens. J.* **2013**, *13*, 4005–4013. [CrossRef]

16. Cinti, S.; Arduini, F.; Moscone, D.; Palleschi, G.; Killard, A.J. Development of a hydrogen peroxide sensor based on screen-printed electrodes modified with inkjet-printed prussian blue nanoparticles. *Sensors* **2014**, *14*, 14222–14234. [CrossRef] [PubMed]

17. Mensing, J.P.; Wisitsoraat, A.; Tuantranont, A.; Kerdcharoen, T. Inkjet-printed sol-gel films containing metal phthalocyanines/porphyrins for opto-electronic nose applications. *Sens. Actuator B* **2013**, *176*, 428–436. [CrossRef]

18. Ferraro, P.; Coppola, S.; Grilli, S.; Paturzo, M.; Vespini, V. Dispensing nano-pico droplets and liquid patterning by pyroelectrodynamic shooting. *Nat. Nanotechnol.* **2010**, *5*, 429. [CrossRef] [PubMed]

19. Coppola, S.; Nasti, G.; Todino, M.; Olivieri, F.; Vespini, V.; Ferraro, P. Direct writing of microfluidic footpaths by pyro-ehd printing. *ACS Appl. Mater. Interf.* **2017**, *9*, 16488–16494. [CrossRef] [PubMed]

20. Mecozzi, L.; Gennari, O.; Rega, R.; Battista, L.; Ferraro, P.; Grilli, S. Simple and rapid bioink jet printing for multiscale cell adhesion islands. *Macromol. Biosci.* **2016**, *16*, 1–6. [CrossRef] [PubMed]

21. Vespini, V.; Coppola, S.; Todino, M.; Paturzo, M.; Bianco, V.; Grilli, S.; Ferraro, P. Forward electrohydrodynamic inkjet printing of optical microlenses on microfluidic devices. *Lab Chip* **2016**, *16*, 326–333. [CrossRef] [PubMed]

22. Hoath, S.D. *Fundamentals of Inkjet Printing*; John Wiley & Sons, Inc.: Hoboken, NJ, USA, 2016.

23. Mogalicherla, A.K.; Lee, S.; Pfeifer, P.; Dittmeyer, R. Drop-on-demand inkjet printing of alumina nanoparticles in rectangular microchannels. *Microfluid. Nanofluid.* **2014**, *16*, 655–666. [CrossRef]

24. Staymates, M.E.; Fletcher, R.; Verkouteren, M.; Staymates, J.L.; Gillen, G. The production of monodisperse explosive particles with piezo-electric inkjet printing technology. *Rev. Sci. Instrum.* **2015**, *86*. [CrossRef] [PubMed]

25. Yoo, H.; Kim, C. Generation of inkjet droplet of non-Newtonian fluid. *Rheol. Acta* **2013**, *52*, 313–325. [CrossRef]

26. Kim, B.-H.; Kim, S.-I.; Lee, J.-C.; Shin, S.-J.; Kim, S.-J. Dynamic characteristics of a piezoelectric driven inkjet printhead fabricated using mems technology. *Sens. Actuators A* **2012**, *173*, 244–253. [CrossRef]

27. Kim, B.-H.; Kim, T.-G.; Lee, T.-K.; Kim, S.; Shin, S.-J.; Kim, S.-J.; Lee, S.-J. Effects of trapped air bubbles on frequency responses of the piezo-driven inkjet printheads and visualization of the bubbles using synchrotron x-ray. *Sens. Actuator A* **2009**, *154*, 132–139. [CrossRef]

28. Kwon, K.-S.; Choi, Y.-S.; Lee, D.-Y.; Kim, J.-S.; Kim, D.-S. Low-cost and high speed monitoring system for a multi-nozzle piezo inkjet head. *Sens. Actuator A* **2012**, *180*, 154–165. [CrossRef]

29. Wijshoff, H. *Structure-and Fluid-Dynamics in Piezo Inkjet Printheads*; University of Twente: Enschede, The Netherlands, 2008.

30. Bogy, D.B.; Talke, F. Experimental and theoretical study of wave propagation phenomena in drop-on-demand ink jet devices. *IBM J. Res. Dev.* **1984**, *28*, 314–321. [CrossRef]

31. Dijksman, J.F. Hydro-acoustics of piezoelectrically driven ink-jet print heads. *Flow Turbul. Combust.* **1998**, *61*, 211–237. [CrossRef]

32. Kim, B.-H.; Lee, H.-S.; Kim, S.-W.; Kang, P.; Park, Y.-S. Hydrodynamic responses of a piezoelectric driven mems inkjet print-head. *Sens. Actuators A* **2014**, *210*, 131–140. [CrossRef]

33. Desai, S.; Lovell, M. Modeling fluid-structure interaction in a direct write manufacturing process. *J. Mater. Process. Technol.* **2012**, *212*, 2031–2040. [CrossRef]

34. Wu, C.H.; Hwang, W.S. The effect of the echo-time of a bipolar pulse waveform on molten metallic droplet formation by squeeze mode piezoelectric inkjet printing. *Microelectron. Reliab.* **2015**, *55*, 630–636. [CrossRef]

35. Liou, T.M.; Chan, C.Y.; Shih, K.C. Effects of actuating waveform, ink property, and nozzle size on piezoelectrically driven inkjet droplets. *Microfluid. Nanofluid.* **2010**, *8*, 575–586. [CrossRef]

36. Stemme, E.; Larsson, S.-G. The piezoelectric capillary injector—A new hydrodynamic method for dot pattern generation. *IEEE Trans. Electron. Devices* **1973**, *20*, 14–19. [CrossRef]

37. Beasley, J. Model for fluid ejection and refill in an impulse drive jet. *Soc. Photogr. Sci. Eng.* **1977**, *21*, 78–82.

38. Dijksman, J. Hydrodynamics of small tubular pumps. *J. Fluid Mech.* **1984**, *139*, 173–191. [CrossRef]

39. Li, K.; Liu, J.; Chen, W.; Ye, L.; Zhang, L. A novel bearing lubricating device based on the piezoelectric micro-jet. *Appl. Sci.* **2016**, *6*. [CrossRef]
40. Jong, J.D.; Jeurissen, R.; Borel, H.; Berg, M.V.D.; Wijshoff, H.; Reinten, H. Entrapped air bubbles in piezo-driven inkjet printing. *Phys. Fluids* **2006**, *18*. [CrossRef]

micromachines

MDPI

Article

Comparative Influences of Fluid and Shell on Modeled Ejection Performance of a Piezoelectric Micro-Jet

Kai Li [1], Jun-kao Liu [1],*, Wei-shan Chen [1] and Lu Zhang [2]

[1] State Key Laboratory of Robotics and System, Harbin Institute of Technology, Harbin 150001, China;
     14b908004@hit.edu.cn or sdcxlikai@126.com (K.L.); cws@hit.edu.cn (W.-s.C.)
[2] Aero Engine Corporation of China (AECC), Harbin Dongan Engine Corporation Limited, Harbin 150001,
     China; zhanglu916hit@163.com
* Correspondence: jkliu@hit.edu.cn; Tel.: +86-451-8641-6119

Academic Editors: Ulrich Schmid and Michael Schneider
Received: 21 November 2016; Accepted: 10 January 2017; Published: 13 January 2017

Abstract: The piezoelectric micro-jet, which can achieve the drop-on-demand requirement, is based on ink-jet technology and small droplets can be ejected out by precise control. The droplets are driven out of the nozzle by the acoustic pressure waves which are generated by the piezoelectric vibrator. The propagation processes of the acoustic pressure waves are affected by the acoustic properties of the fluid and the shell material of the micro-jet, as well as the excitations and the structure sizes. The influences of the fluid density and acoustic velocity in the fluid on the nozzle pressure and support reaction force of the vibrator are analyzed in this paper. The effects of the shell material on the ejection performance are studied as well. In order to improve the ejection performance of the micro-jet, for ejecting a given fluid, the recommended methods of selecting the shell material and adjusting excitations are provided based on the results, and the influences of the factors on working frequencies are obtained as well.

Keywords: piezoelectric micro-jet; ejection performance; modelling and simulation

1. Introduction

The working process of the piezoelectric micro-jet is based on the inverse piezoelectric effect and the core component is the piezoelectric vibrator. Acoustic pressure waves are generated in the fluid when pulse voltages are applied on the piezoelectric material, and then the droplets are ejected out of the nozzle by the pressure waves. The piezoelectric micro-jet technology, which can realize drop-on-demand, is widely used in electric and electronic fields [1,2], the bearing lubrication field [3], the manufacturing field [4–7], and the biomedical field [8,9], thanks to its advantages of fast response, good stability, simple structure, etc.

The influences of the pulse voltage parameters on the ejection performance of the micro-jet are widely studied by researchers based on the analysis of the hydrodynamic characteristics [10–14]. However, the ejection performance of the micro-jet is also affected by the acoustic properties of the fluid and the acoustic impedance of the shell. The propagation processes of the acoustic waves in the cavity are affected by the properties of the fluid, and the reflection/absorption processes of the acoustic waves, mainly at the interface between the fluid and shell, are related to the acoustic impedance of the shell [15].

The influences of the fluid acoustic properties and acoustic impedance of the shell on the ejection performance are analyzed in this paper. Based on the analysis results, the methods of shell material selection and excitation adjustment are given, when ejecting fluid with different properties, so as to ensure the ejection intensity of the micro-jet. In order to ensure the restraint stiffness of the vibrator, the reaction forces of the vibrator under different conditions are studied as well.

As for different applications, the designed structure of the piezoelectric micro-jet are different. The results shown in this paper are mainly used for the given common structure style shown in the text. For other structures, results can be obtained according to the proposed analysis simulation method.

2. Structure and Methods

A piezoelectric micro-jet, which is simple and common, as shown in Figure 1a, is studied in this paper. The piezoelectric vibrator is made by gluing the piezoelectric ceramic (PZT) on the copper diaphragm. When pulse voltages are applied on the piezoelectric vibrator, the vibrator vibrates and acoustic pressure waves are created in the fluid domain. Then, the droplets are pushed out from the nozzle by the acoustic pressure waves. The structure sizes of the piezoelectric micro-jet are shown in Figure 1b. The thickness of the piezoelectric ceramic and copper diaphragm used in this paper are 0.4 mm and 0.2 mm, respectively.

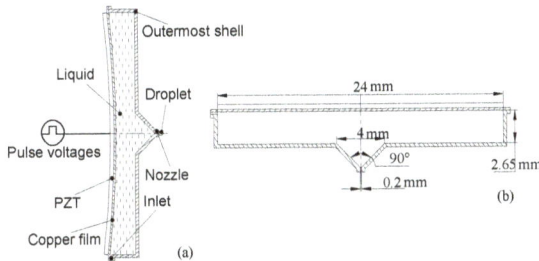

Figure 1. The structure and structure sizes of the piezoelectric: (**a**) the structure of the traditional piezoelectric micro-jet; and (**b**) the structure sizes of the micro-jet.

The acoustic pressure distribution in the fluid field of the piezoelectric micro-jet, after the vibration of the piezoelectric vibrator, is shown in Figure 2. As we can see, due to the transmission loss in the fluid domain and the reflection/absorption process at the interface between the fluid and shell, the acoustic pressure distribution in the cavity is not uniform, and the pressure level decreases gradually from the interface to the nozzle. As the pressure distribution is difficult to measure with sensors, it is difficult to study the acoustic pressure characteristics of the micro-jet through experiments. Thus, the correlation analyses are carried out by finite element analysis with simulation software ANSYS Workbench (Version 14.5, ANSYS, Inc., Canonsburg, PA, USA) which has been installed ExtAcoustics, ExtPiezo, and FSI_transient extensions.

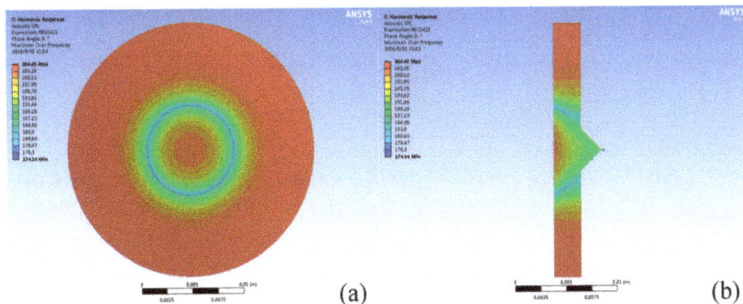

Figure 2. The acoustic pressure distribution in the fluid field of the piezoelectric micro-jet: (**a**) the global state of the pressure distribution; and (**b**) the cross-section diagram of the pressure distribution.

3. Boundary Conditions

The model used in simulations is shown in Figure 3, we can see that the inlet part of the micro-jet is neglected due to its small size, and the sweep method is selected as the meshing method to improve the accuracy of calculations. As the effects of shell on acoustic pressure waves are determined by the acoustic impedance of the shell, the impedance boundary conditions are applied on the surfaces of the fluid domain instead of using the shell element.

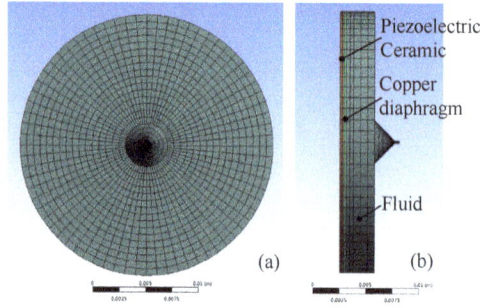

Figure 3. The model for simulation: (**a**) the front view of the model; (**b**) the left view of the model.

The type of the piezoelectric ceramic is selected as PZT-5H, and its physical properties (elastic stiffness constant matrix $[c^E]$, piezoelectric stress constant matrix $[e]$, and dielectric constant matrix $[\varepsilon^T]$) in the simulation are set as:

$$[c^E] = \begin{bmatrix} 13.2 & 7.3 & 7.1 & 0 & 0 & 0 \\ 7.3 & 13.2 & 7.1 & 0 & 0 & 0 \\ 7.1 & 7.1 & 11.5 & 0 & 0 & 0 \\ 0 & 0 & 0 & 2.6 & 0 & 0 \\ 0 & 0 & 0 & 0 & 2.6 & 0 \\ 0 & 0 & 0 & 0 & 0 & 3 \end{bmatrix} \times 10^{10} (N/m^2)$$

$$[e] = \begin{bmatrix} 0 & 0 & -2.4 \\ 0 & 0 & -2.4 \\ 0 & 0 & 17.3 \\ 0 & 0 & 0 \\ 0 & 12.95 & 0 \\ 12.95 & 0 & 0 \end{bmatrix} (C/m^2)$$

$$[\varepsilon^T] = \begin{bmatrix} 804.6 & 0 & 0 \\ 0 & 804.6 & 0 \\ 0 & 0 & 659.7 \end{bmatrix} \times 10^{-11} (F/m)$$

The elasticity modulus, density, and Poisson ratio of the copper film are set as 7.65×10^{10} N/m^2, 7.5×10^3 kg/m^3, and 0.32, respectively. The outer ring of the copper diagram is set as a fixed constraint. The nozzle is the interface between fluid and air; therefore, the acoustic impedance at the nozzle is set as 400 kg·m^{-2}·s^{-1} according to the related data [16]. In order to obtain the influences of the acoustic impedance of shell on the ejection performance of the micro-jet, the shell materials with different impedance are selected and the impedance values of different materials are shown in Table 1. The pulse voltages are applied to the surface of the piezoelectric ceramic and the amplitudes of the voltages are set as 200 V. As structure-acoustic coupling is created at the interface between the fluid domain and the piezoelectric vibrator, the boundary condition at the interface is set as a fluid-solid interface.

Table 1. Acoustic impedance of different materials ($\times 10^6$ Kg·m^{-2}·s^{-1}).

Material	Plexiglass	Magnesium	Aluminum	Iron	Copper
Impedance	3.1	10.0	17.1	33.2	41.6

4. Results and Discussion

When the cavity of the micro-jet is filled with liquid, the vibration of the vibrator will be subjected to the reaction force of the fluid at the fluid-solid interface due to the fluid-solid coupling effect. The obvious influence of the fluid-solid coupling effect on the vibration of the vibrator is that the resonant frequencies of the vibrator are all reduced.

When the cavity is filled with air, the mode shapes and the corresponding resonant frequencies of the first four order modes of the vibrator are shown in Figure 4, we can see that in terms of the structure proposed in this paper, the first-order modal and the fourth-order mode are more suitable, as the nozzle of the piezoelectric micro-jet is located in the middle.

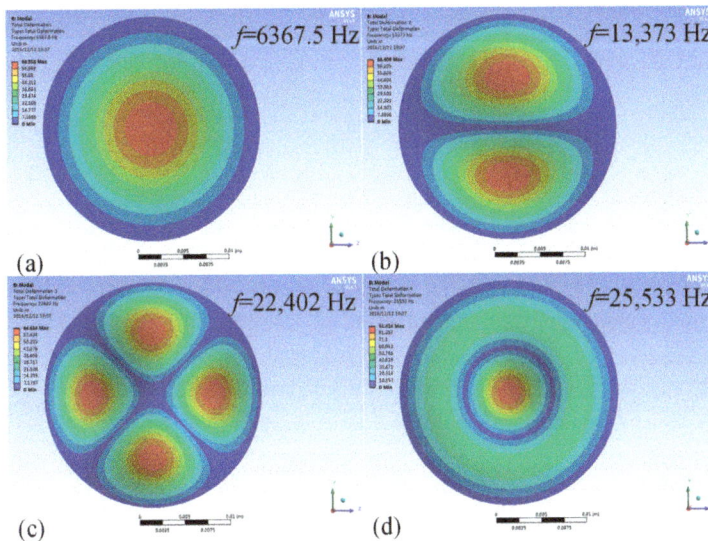

Figure 4. The mode shapes and the corresponding resonant frequencies of the vibrator when the cavity is filled with air: (**a**) first-order mode with the resonant frequency as 6367.5 Hz; (**b**) second-order mode with the resonant frequency as 13,373 Hz; (**c**) third-order mode with the resonant frequency at 22,402 Hz; and (**d**) fourth-order mode with resonant frequency at 25,533 Hz.

When the cavity is filled with fluid, the mode shapes and the corresponding resonant frequencies are shown in Figure 5, and it can be seen that the resonant frequencies of the vibrator are all reduced significantly due to the fluid-solid coupling effect, while the mode shapes remain unchanged. The first-order working mode is suitable for low-frequency intermittent ejection, and the fourth-order mode is suitable for high-frequency continuous ejection.

As the constraint stiffness of the vibrator should meet the requirements, which are determined by the reaction force of the vibrator, so as to ensure the stability of the vibrator, the frequency response curves of the reaction force under different conditions are obtained. The frequency response characteristics of nozzle pressure are analyzed, as the ejection intensity is reflected by the acoustic pressure at the nozzle. To obtain better ejection performance, the vibration frequency of the vibrator, which determines the working frequency of the micro-jet, should be consistent with its resonance frequency. Thus, the influences of fluid properties and shell material on the resonant frequency of the vibrator are analyzed as well.

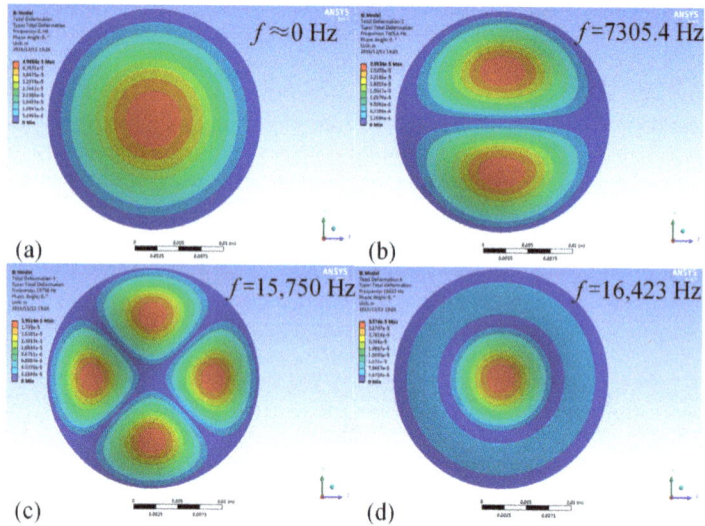

(a) $f \approx 0$ Hz (b) $f = 7305.4$ Hz
(c) $f = 15,750$ Hz (d) $f = 16,423$ Hz

Figure 5. The mode shapes and the corresponding resonant frequencies of the vibrator when the cavity is filled with fluid: (**a**) first-order mode with the resonant frequency as close to 0 Hz; (**b**) second-order mode with the resonant frequency as 7305.4 Hz; (**c**) third-order mode with the resonant frequency as 15,750 Hz; and (**d**) fourth-order mode with the resonant frequency as 16,423 Hz.

4.1. Influences of the Density of Fluid

The frequency response curves of the nozzle pressure of the micro-jet, when the acoustic velocity in the fluid is set as 1400 m/s and the material of the shell is set as aluminum, and when the fluids with different densities are ejected, are shown in Figure 6. As we can see that there are two peaks in each frequency response curve, which occur at the frequencies corresponding to the first-order and fourth-order resonant frequencies of the vibrator. It also demonstrates that the nozzle pressure of the micro-jet is less affected by the density of the fluids when the vibrator vibrates in the first-order mode.

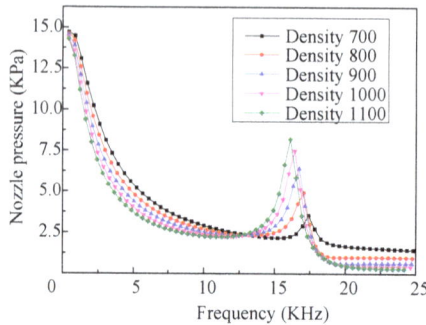

Figure 6. The frequency response curves of the nozzle pressure of the micro-jet (fluid with different densities).

As the micro-jet generally operates at high frequency, so as to obtain high working efficiency, only the influences of the density on the nozzle pressure of the micro-jet, when the vibrator vibrates in the fourth-order mode, are analyzed. The effects of the densities on the amplitude of the nozzle pressure, when the vibrator vibrates in the fourth-order mode, are shown in Figure 7. We can see that with the increase of density, the amplitude of the nozzle pressure increases gradually, and the change trend is close to linear.

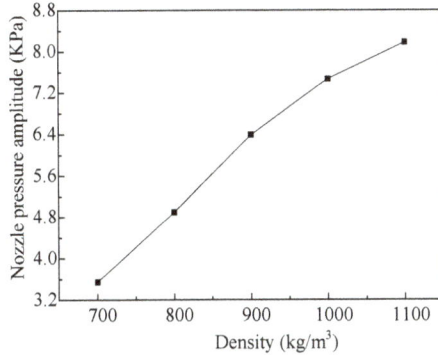

Figure 7. The amplitudes of the nozzle pressure vary with different densities of the fluid.

The response curves of the reaction force of the vibrator, when fluids with different densities are ejected, are shown in Figure 8. We can see that there is only on peak value in each response curve of the reaction force of the vibrator and they all occur at the frequencies corresponding to the fourth-order mode. Additionally, the reaction force of the vibrator, when it vibrates in the first-order mode, is very small. The reason is that the inertia force created by the vibration of the vibrator, which determines the reaction force, relates to the vibration velocity of the vibrator, and the vibration velocity of the vibrator working in the fourth-order mode is larger than that of working in the first-order mode. Thus, the required restraint stiffness when the vibrator works in the fourth-order mode should be larger.

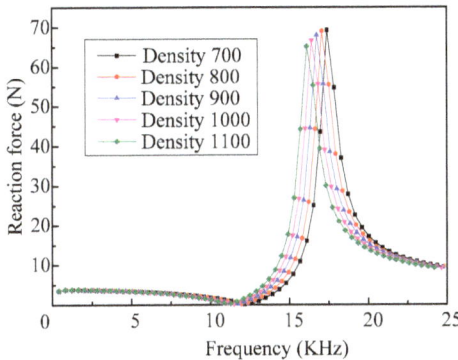

Figure 8. The frequency response curves of the reaction force of the vibrator when ejecting fluids with different densities.

As we can see from Figure 8, when the vibrator vibrates in the first-order mode, density has little effect on the reaction force of the vibrator. Thus, only the influences of the density, when the vibrator works in the fourth-order mode, on the ejection performance are analyzed here. The relationship between the reaction force amplitude of the vibrator which working in the fourth-order mode, and the densities of the fluids to be ejected is shown in Figure 9. As we can see, with the increase of the density, the amplitude of the reaction force decreases gradually, and the changes of amplitude increases.

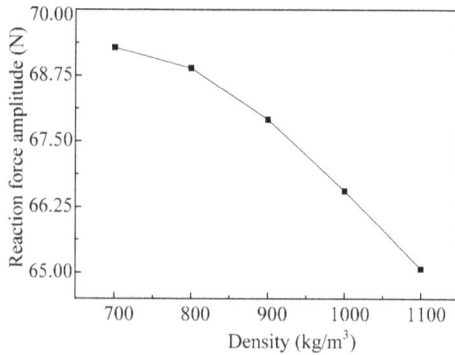

Figure 9. The amplitudes of the reaction force vary with different densities of the fluids.

As we can see from Figure 6, when ejecting fluid with different densities, the resonant frequencies of the vibrator working in the first-order mode are different, but the difference is small, and the influences of the density on the resonant frequency of the vibrator working in the fourth-order mode are shown in Figure 10. We can see that the resonance frequency of the vibrator decreases nearly linearly with the increase of the density. Thus, the operating frequency of the micro-jet decreases when the density of the fluid increases.

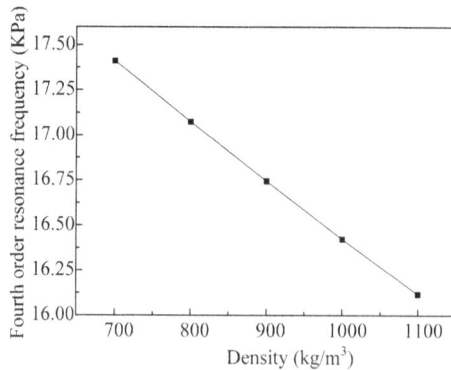

Figure 10. The fourth-order resonance frequencies of the vibrator varies with different fluid densities.

4.2. Influences of the Acoustic Velocity in Fluid

When the density of the fluid is set as 1000 kg/m^3, the material of the shell is set as aluminum, and the acoustic velocity in the fluid is set differently, the frequency response curves of the nozzle pressure are shown in Figure 11. As we can see, there are also two peaks that occur in each frequency response curve, which correspond to the first-order and fourth-order modes. It can be seen that the frequency response curves of the nozzle pressure, when ejecting fluid with different acoustic velocity, is nearly the same. Therefore, when the micro-jet works at low frequency, the effect of acoustic velocity in the fluid on the nozzle pressure can be ignored.

As the acoustic velocity in the fluid has little influence on the nozzle pressure when the micro-jet is working at low frequency, only the relationships between the acoustic velocity and the nozzle pressure when the micro-jet is working at high frequency are obtained. The curve of the nozzle pressure amplitude along with the increase of the acoustic velocity of the fluid is shown in Figure 12. We can see that with the increase of the acoustic velocity, the amplitude of the nozzle pressure increases gradually and the change trend is close to linear.

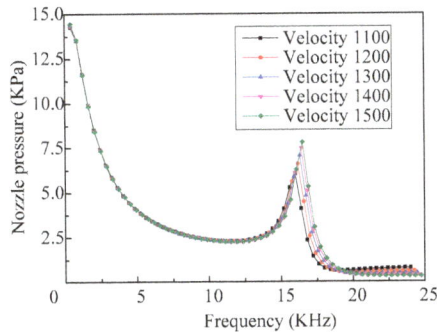

Figure 11. The frequency response curves of the nozzle pressure of the micro-jet (fluid with different acoustic velocity).

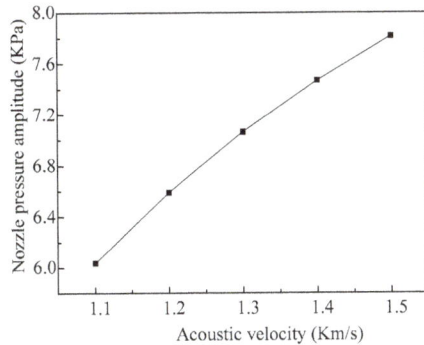

Figure 12. The amplitudes of the nozzle pressure vary with different acoustic velocity of the fluids.

The response curves of the reaction force of the vibrator, when fluids with different acoustic velocity are ejected, are shown in Figure 13. We can see that the effect of acoustic velocity on the reaction force of the vibrator, when it vibrates at low frequency, is also very small.

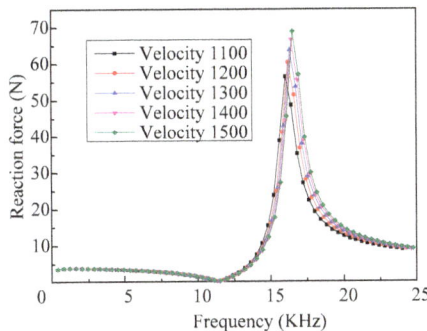

Figure 13. The frequency response curves of the reaction force of the vibrator when ejecting fluids with different acoustic velocities.

The relationship between the reaction force amplitude of the vibrator working in the fourth-order mode and the acoustic velocity in the fluid is shown in Figure 14. As we can see, with the increase of the acoustic velocity, the amplitude of the reaction force increases gradually, and the changes of the amplitude decreases.

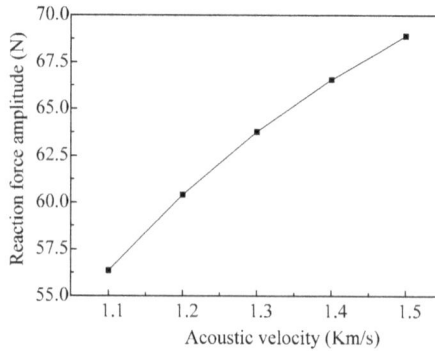

Figure 14. The amplitudes of the reaction force vary with different acoustic velocities of the fluid.

When the vibrator vibrates in the fourth-order mode, the influences of the fluid acoustic velocity on the resonant frequency of the vibrator are shown in Figure 15. We can see that the resonance frequency of the vibrator increases gradually with the increase of the acoustic velocity of the fluid, and the changes of the frequency decrease. Therefore, the operating frequency of the micro-jet increases along with the increase of the acoustic velocity of the fluid.

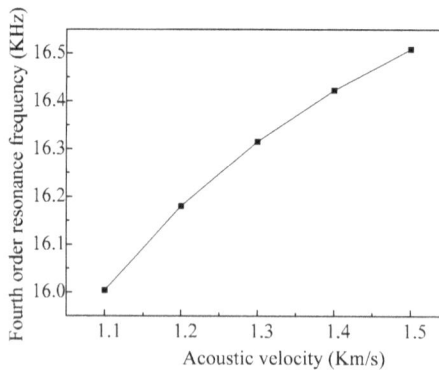

Figure 15. The fourth-order resonance frequencies of the vibrator varies with different fluid densities.

4.3. Influences of the Shell Material

When selecting different shell materials and the density and acoustic velocity of the fluid are set as 1000 kg/m^3 and 1400 m/s, respectively, the frequency response curves of the nozzle pressure of the micro-jet are shown in Figure 16. As we can see, the impedances of the shell materials have little influence on the first-order and fourth-order resonant frequencies of the vibrator, whereas, with the changes of the impedance of the shell material, the changes of the peak value of the nozzle pressure are relatively large.

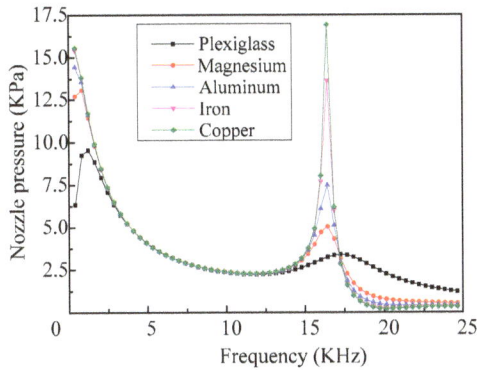

Figure 16. The frequency response curves of the nozzle pressure of the micro-jet (shell with different materials).

The relationship curves between the material impedance of the shell and the nozzle pressure, when the vibrator works in the first-order mode and fourth-order mode, are shown in Figure 17. It can be seen that when the acoustic impedance of the shell material is less than 36×10^6 kg·m^{-2}·s^{-1}, the nozzle pressure amplitudes of the micro-jet when working in the first-order mode are greater than that of working in the fourth-order mode. With the increase of the acoustic impedance of the shell, the nozzle pressure amplitude increases gradually and the change tends to be stable when the micro-jet is working in the first-order mode. The nozzle pressure amplitude of the micro-jet, working in the fourth-order mode, also increases along with the increase of the acoustic impedance of the shell; however, the change tends to be enhanced.

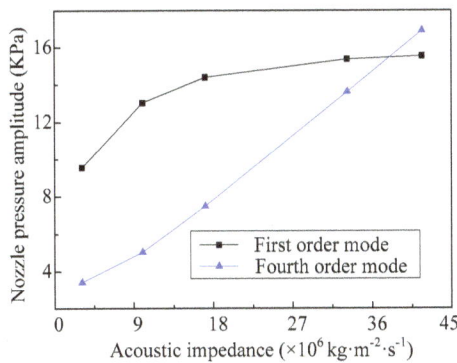

Figure 17. The amplitudes of the nozzle pressure vary with different acoustic impedance of the shell.

When the shell materials with different acoustic impedances are selected, the response curves of the reaction force of the vibrator are shown in Figure 18. We can see that when the micro-jet works at a low frequency that the reaction force of the vibrator is little affected by the impedance of the shell material. However, the impedance of the shell material has a great influence on the reaction force of the vibrator, when the vibrator works in the fourth-order mode.

The influences of the impedance of the shell material on the reaction force amplitude of the vibrator, when the vibrator works in the fourth-order mode, are shown in Figure 19. We can see that the reaction force amplitude of the vibrator increases near linearly with the increase of the acoustic impedance of the shell.

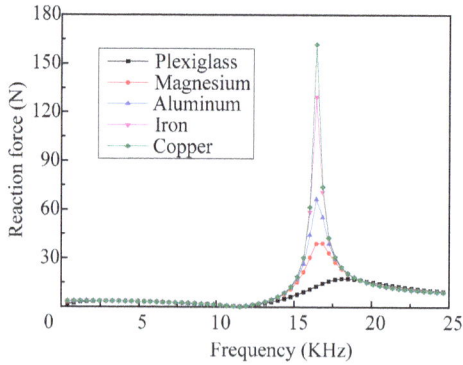

Figure 18. The frequency response curves of the reaction force of the vibrator when selecting shells with different acoustic impedance.

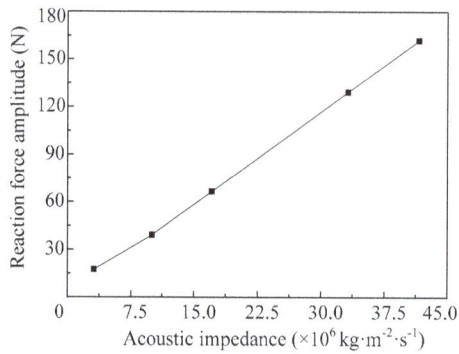

Figure 19. The amplitudes of the reaction force vary with different acoustic impedance of the shell.

5. Conclusions

When fluid with relatively low density and low acoustic velocity is ejected, the shell material with high acoustic impedance should be selected and the pulse voltage excitation should be enhanced, so as to ensure the strength of the ejection.

In order to guarantee the required restraint stiffness and stability of the vibrator, when fluid with relatively low density and high acoustic velocity is ejected, the shell material with small acoustic impedance is recommended to reduce the reaction force amplitude of the vibrator.

Ejecting fluid with relatively small density and high acoustic velocity, when there are a variety of fluids that can be selected, can increase the working frequency of the micro-jet, which means the working efficiency of the micro-jet can be improved. In addition, the influences of the acoustic impedance of the shell on the operating frequency of the micro-jet can be neglected.

Acknowledgments: This project is supported by the National Natural Science Foundation of China (Grant Nos. 51075082 and 51375107).

Author Contributions: Kai Li and Jun-kao Liu designed and performed the simulations; Wei-shan Chen, Kai Li and Lu Zhang analyzed the data; Wei-shan Chen contributed analysis tools; Kai Li and Lu Zhang wrote the paper.

Conflicts of Interest: The authors declare no conflict of interest.

References

1. Kwon, Y.T.; Lee, Y.I.; Lee, K.J.; Choi, Y.M.; Choa, Y.H. A novel method for fine patterning by piezoelectrically induced pressure adjustment of inkjet printing. *J. Electron. Mater.* **2015**, *44*, 2608–2614. [CrossRef]
2. Sielmann, C.J.; Busch, J.R.; Stoeber, B.; Walus, K. Inkjet printed all-polymer flexural plate wave sensors. *IEEE Sens. J.* **2013**, *13*, 4005–4013. [CrossRef]
3. Li, K.; Liu, J.-K.; Chen, W.-S.; Zhang, L. Effects of pulse voltage on piezoelectric micro-jet for lubrication. *Microsyst. Technol.* **2016**, 1–9, in press. [CrossRef]
4. Torabi, P.; Petros, M.; Khoshnevis, B. Enhancing the resolution of selective inhibition sintering (SIS) for metallic part fabrication. *Rapid Prototyp. J.* **2015**, *21*, 186–192. [CrossRef]
5. Tenda, K.; Ota, R.; Yamada, K.; Henares, T.G.; Suzuki, K.; Citterio, D. High-resolution microfluidic paper-based analytical devices for sub-microliter sample analysis. *Micromachines* **2016**, *7*, 80. [CrossRef]
6. Wang, H.-L.; Chu, C.-H.; Tsai, S.-J.; Yang, R.-J. Aspartate aminotransferase and alanine aminotransferase detection on paper-based analytical devices with inkjet printer-sprayed reagentsl. *Micromachines* **2016**, *7*, 9. [CrossRef]
7. Jacot-Descombes, L.; Gullo, M.R.; Cadarso, V.J.; Mastrangeli, M.; Ergeneman, O.; Peters, C.; Fatio, P.; Freidy, M.A.; Hierold, C.; Nelson, B.J.; et al. Inkjet printing of high aspect ratio superparamagnetic SU-8 microstructures with preferential magnetic directions. *Micromachines* **2014**, *5*, 583–593. [CrossRef]
8. Lorber, B.; Hsiao, W.K.; Hutchings, I.M.; Martin, K.R. Adult rat retinal ganglion cells and glia can be printed by piezoelectric inkjet printing. *Biofabrication* **2014**, *6*, 9. [CrossRef] [PubMed]
9. Gross, A.; Schondube, J.; Niekrawitz, S.; Streule, W.; Riegger, L.; Zengerle, R.; Koltay, P. Single-cell printer: Automated, on demand, and label free. *J. Lab. Autom.* **2013**, *18*, 504–518. [CrossRef] [PubMed]
10. Li, K.; Liu, J.; Chen, W.; Ye, L.; Zhang, L. A novel bearing lubricating device based on the piezoelectric micro-jet. *Appl. Sci.* **2016**, *6*, 38. [CrossRef]
11. Kim, B.H.; Kim, S.I.; Shin, H.H.; Park, N.R.; Lee, H.S.; Kang, C.S.; Shin, S.J.; Kim, S.J. A study of the jetting failure for self-detected piezoelectric inkjet printheads. *IEEE Sens. J.* **2011**, *11*, 3451–3456. [CrossRef]
12. Wu, H.C.; Hwang, W.S.; Lin, H.J. Development of a three-dimensional simulation system for micro-inkjet and its experimental verification. *Mater. Sci. Eng. A* **2004**, *373*, 268–278. [CrossRef]
13. Liou, T.M.; Chan, C.Y.; Shih, K.C. Study of the characteristics of polymer droplet deposition in fabricated rectangular microcavities. *J. Micromech. Microeng.* **2009**, *19*, 12. [CrossRef]
14. Li, K.; Liu, J.-K.; Chen, W.-S.; Ye, L.; Zhang, L. Research on the injection performance of a novel lubricating device based on piezoelectric micro-jet technology. *J. Electron. Mater.* **2016**, *45*, 4380–4389. [CrossRef]
15. Derby, B. Inkjet printing ceramics: From drops to solid. *J. Eur. Ceram. Soc.* **2011**, *31*, 2543–2550. [CrossRef]
16. Acoustic Impedance. Available online: http://radiopaedia.Org/articles/acoustic-impedance (accessed on 5 August 2016).

micromachines

MDPI

Article

A PZT Actuated Triple-Finger Gripper for Multi-Target Micromanipulation

Tao Chen, Yaqiong Wang, Zhan Yang, Huicong Liu *, Jinyong Liu and Lining Sun

Jiangsu Provincial Key Laboratory of Advanced Robotics & Collaborative Innovation Center of Suzhou Nano Science and Technology, Soochow University, Suzhou 215123, China; chent@suda.edu.cn (T.C.); 20154229003@stu.suda.edu.cn (Y.W.); yangzhan@suda.edu.cn (Z.Y.); 20134229002@stu.suda.edu.cn (J.L.); lnsun@hit.edu.cn (L.S.)
* Correspondence: hcliu078@suda.edu.cn; Tel.: +86-512-6758-7217

Academic Editors: Ulrich Schmid and Michael Schneider
Received: 25 November 2016; Accepted: 16 January 2017; Published: 24 January 2017

Abstract: This paper presents a triple-finger gripper driven by a piezoceramic (PZT) transducer for multi-target micromanipulation. The gripper consists of three fingers assembled on adjustable pedestals with flexible hinges for a large adjustable range. Each finger has a PZT actuator, an amplifying structure, and a changeable end effector. The moving trajectories of single and double fingers were calculated and finite element analyses were performed to verify the reliability of the structures. In the gripping experiment, various end effectors of the fingers such as tungsten probes and fibers were tested, and different micro-objects such as glass hollow spheres and iron spheres with diameters ranging from 10 to 800 μm were picked and released. The output resolution is 145 nm/V, and the driven displacement range of the gripper is 43.4 μm. The PZT actuated triple-finger gripper has superior adaptability, high efficiency, and a low cost.

Keywords: micromanipulation; triple-finger; micro-gripper

---

1. Introduction

Applications of micro-manipulation can be extensively found in microsystem manufacture, micro-medical, biomedical, optical engineering, and other important areas [1]. An increased requirement in micro-manipulation has led to rapid development of manipulating grippers. For micro/nano-manipulation, the manipulated objects usually have a small size, typically ranging from 1 μm to 1 mm, and are difficult to handle. Hence, various micro-grippers have been designed and developed based on different actuation techniques.

Grippers in the form of double- or triple-finger are normally driven by magnetic, electrostatic, or piezoelectric actuation mechanisms [2]. The implementation of magnetic actuator requires a strict manipulating environment, which cannot be affected by the ambient magnetic field interaction. Due to the disadvantageous scaling of magnetic fields, magnetic microactuators have low force output which limited their performance. The electrostatic actuation mechanism based on micro-electromechanical systems (MEMS) can realize high precision with a small device size and generate a satisfactory amount of output force [3,4]. But, the integrated structure is fragile and the micro-fabrication process is complicated. Xu et al. developed a micro-gripper with integrated electrostatic actuator and capacitive force sensor. The feasibility and effectiveness of the gripper device were validated by gripping a human hair 76 μm of diameter [5]. Boudaoud et al. fabricated an electrostatic micro-gripper with a nonlinear actuation mechanism while handling calibrated micro glass balls 80 μm of diameter [6]. Brandon et al. made a MEMS gripper which can pick and place gold nano spheres. The gripper was able to withstand gripping forces more than 700 μN and 38 μm out-of-plane bending deflections, proving the gripping tips to be mechanically strong for micro/nano manipulation [7]. The pick-and-place of a 100 nm gold nano sphere inside an SEM has been demonstrated with this gripper. Due to limitation of micro-fabrication process, the manipulation range of these two electrostatic grippers cannot be adjusted. Piezoelectric micro-actuators, with their high speed and good motion resolution, have been widely advocated to implement high precision grasping manipulation [8]. A major disadvantage of a piezoelectric actuation mechanism is that the output motion is relatively small (typically about 10–20 μm). To solve this problem, displacement amplification mechanisms are usually employed [9]. Wang et al. developed a monolithic compliant piezoelectric-driven micro-gripper with a large displacement amplification ratio of 16 [10].

Grippers with a wide clamping range are adaptable to manipulate various objects. Thus, the manipulation efficiency can be greatly improved. Sun et al. developed a piezo-actuated flexure-based micro-gripper being capable of handling various sized micro-objects with a maximum jaw displacement of 150.8 μm and a high amplification ratio of 16.4 [11]. It adopted monolithic structure and had a fixed manipulating range. Nevertheless, most structures of the reported micro-grippers cannot be changed according to different targets, thus confining the jaw displacement to variously sized micro-objects. Compared with the monolithic structure of a micro-gripper, a multi-actuator system driven by numerous identical single actuators connected in parallel and in series is highly adaptable to handle different objects. In this light, it is the best choice for the mechanical structure design of the micro-gripper.

A common design attribute of the micro-grippers proposed in recent years is a double-finger design [12], which has been proved to have a better performance in grasping micro-objects [13,14]. However, it could not release objects effectively due to the adhesion force [15]. In the micro scale, the strong adhesion force would make the micro-object adhere to the end-effector during release, especially for the objects under 50 μm. To solve this issue, many researchers have conducted significant work on the approaches to ease the difficulty of release. Those solutions and manipulation strategies can be roughly divided into two types, passive release and active release.

The basic idea of passive release is to increase the adhesion force between object and substrate or decrease the adhesion force between object and gripper. Liquid with great viscosity was painted on the basement to increase the adhesion force between object and substrate [16]. By scrolling the object, contacting surface can be reduced to achieve a success release [17]. These strategies are strongly dependent on the surface properties of substrate, and are greatly influenced by the materials and environment. Each process of release would take a very long time with low efficiency and poor operation repeatability.

Other approaches for the separation of objects and manipulators are proposed by using an external electric field, magnetic field force [18] or liquid frozen method [19], which called active release without contacting with the substrate. Those kinds of releasing ways do not depend on the surface properties of the substrate, but have difficulty in the miniaturization of grippers and require complex peripherals. Yet the release precision is poor. Sun et al. presented a novel MEMS gripper with a plunging structure for releasing the objects adhered to a gripping arm [20]. Although these triple-arm grippers are able to releasing the microspheres, the temporary impact of the plunging arm is so strong that the structure of some soft micro-objects may be damaged. And, the tilt thrust may affect the release accuracy to a certain extent. High frequency mechanical vibration of manipulation tools for object releasing are also conducted though piezoelectric ceramics driving [21], while the precise positioning is not achieved.

This paper reports the design of a separate-structured triple-finger gripper with a piezoelectric actuation mechanism for multi-target micromanipulation. Different from those grippers with gripping arms or fingers distributed in the same plan, this triple-finger gripper are designed in a spatial distributed structure enabling highly stable and repeatable pick-and-place of micro-objects. Grippers described above are limited in their clamping range. By adjusting the positions of the three separate assembly modules, which can be replaced by a variety of components suitable for different tasks, the gripper is able to achieve a wide clamping range and a superior adaptability for objects in different size. It would be less costly for manipulating multi-targets with this type of gripper compare to those grippers limited in a determinate range. In addition, the end effectors of three fingers can be interchangeable in shape, material, and size according to different shaped manipulating targets. Moreover, the third finger aided for releasing changed the acceleration of the releasing object by knocking on the object. The releasing process aided with a third finger overcomes the adhesion force between the third finger and the objects, which achieves an effective release for objects. In Section 1, we have given a general talk about some previous research on micro-grippers. The detailed structure and motion analysis of the proposed gripper are described in Section 2. Simulations and Calibration of the gripper are presented in Section 3. Experimental results of pick-and-place micro-objects by double-finger or triple-finger gripper and discussions are included in Section 4.

2. Design and Motion Analysis of Gripper

Figure 1a shows a schematic diagram of the micro-gripper with a triple-finger. The overall size of the gripper is about 107 mm × 94 mm × 67 mm. The left, right, and the up fingers are assembled in an *x-y*-axis adjustable pedestal via three adjustable blocks. Each finger contains a *z*-axis adjustable pedestal, an amplifying structure, and an end-effector. The amplifying structure is driven by a piezoceramic (PZT) actuator which can generate a maximum displacement of 10 μm. The structure of the *z*-axis adjustable pedestal is specially designed with five pairs of thread holes on the both sides of the groove to adjust the initial clamping range of the gripper. By fixing the fingers on the different holes, the gripper can reach a wide clamping range to meet the requirements of micro-objects with different sizes and shapes. According to different manipulating micro-objects, variable end effectors can be replaced, such as tungsten tipped probes, fibers, atomic force microscope (AFM, Olympus, Tokyo, Japan) probes and so on.

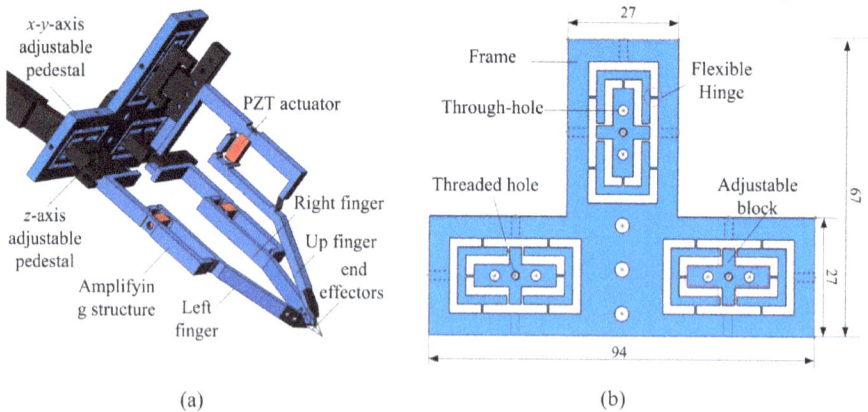

(a)   (b)

Figure 1. (**a**) Schematic drawing of the triple-finger gripper; (**b**) Front view of the *x-y* adjustable pedestal.

The *x-y*-axis and *z*-axis adjustable pedestals are employed to expand the range of the clamping objects and avoid deviation in a certain range. Figure 1b shows a front view of the *x-y*-axis adjustable pedestal. It contains three adjustable blocks. Two through-holes and one threaded-hole in each adjustable block are used to connect and modulate the *z*-axis adjustable pedestal. Flexible hinges are designed in between the adjustable blocks and the frame of the pedestal. By using swivel bolts through the adjustable holes in the frame, every block is able to make micrometric displacements along *x*- and *y*-axis. Figure 2a shows a simplified diagram of the *x-y*-axis motion of the blocks. The frame is assumed as the fixed end and the flexible hinges are simplified as cross lines. Every adjustable block is driven by three swivel bolts along *x* and *y*-directions as identified in red arrow.

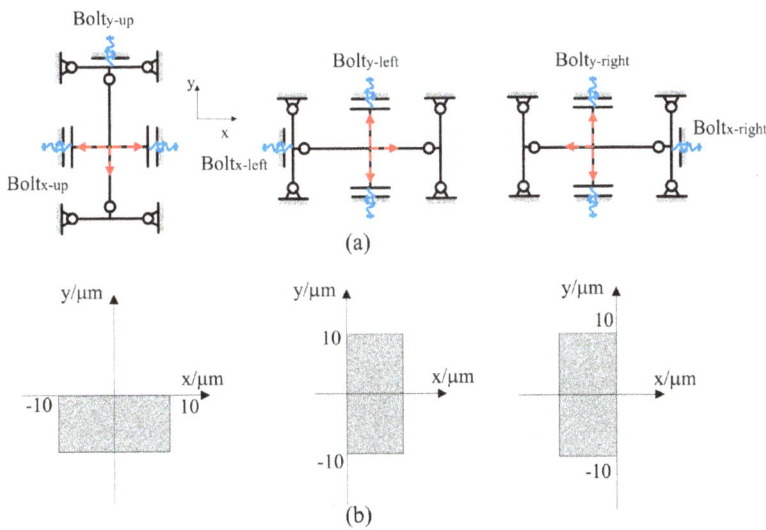

Figure 2. (a) Motion diagram of the *x-y* adjustable pedestal; (b) The displacements of the up, left, and right blocks in *x-y* plane, respectively.

Taking the center of every adjustable block as the datum point, the displacement range of every block is calculated. As shown in Figure 2b, the movement range of the up block is from -10 to 10 μm in *x*-axis and from -10 to 0 μm in *y*-axis. The displacement range of the left block is from 0 to 10 μm in *x*-axis and from -10 to 10 μm in *y*-axis. The range of the right block is from -10 to 0 μm in *x*-axis and from -10 to 10 μm in *y*-axis. With these adjustable blocks, three separate fingers can be aligned accurately. With a resolution of micrometers scale, the gripper can be assembled in a relatively high precision for clamping micro-objects.

Figure 3a shows the assembled single finger structure and Figure 3b shows the simplified motion diagram of a single finger. The bolt can adjust the displacement of the finger along *z*-axis direction in a certain range. The amplification mechanism is essentially a micro-leverage structure, which contains two flexible hinges, a spring structure, and an embedded PZT actuator. The two flexible hinges can be considered as a fulcrum point. The PZT actuator is pre-tightened initially by a bolt and it produces a driving force F on the amplfying structure at point A. The spring structure helps the finger to recover the original position as the driving force of the PZT transducer is removed.

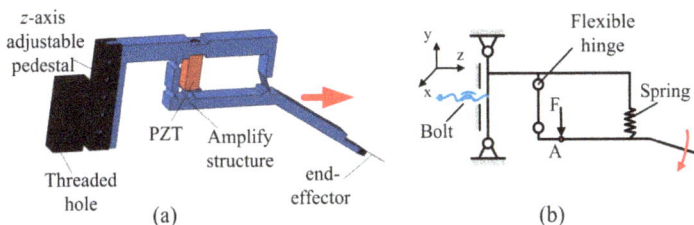

Figure 3. Assembled single finger structure and motion diagram of a single finger.

Figure 4a shows the dimension parameters of a simplified single finger. θ_1 is the angle between OB_1 and OC_1. *a* is the length of OD, i.e., 62.5 mm, and *d* is the length of B_1D, i.e., 30 mm. *c* is the length of OC_1.

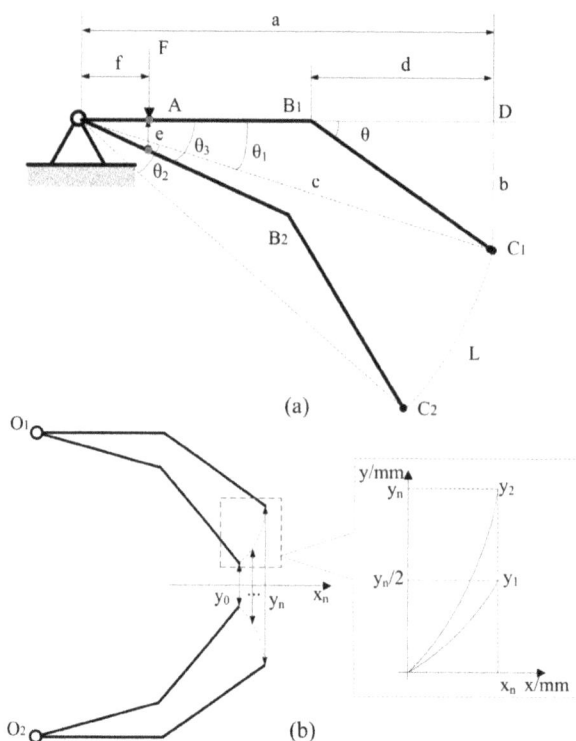

Figure 4. (a) Simplified diagram and the maximum output of a single finger (unit: mm); (b) Trajectories of left and right fingers and gripping distance between these two fingers.

Assuming the PZT actuator applied a driving force F at point A, the output displacement of point A on the finger is e. The moving angle can be expressed as:

$$\tan \theta_3 = e/F \tag{1}$$

The maximum output displacement of piezoelectric transducer is 0.01 mm, which is the length of e. θ_2 is the moving angle of the finger, which equals to θ_3. Trajectory length L of the single finger is calculated to be 0.118 mm.

Figure 4b shows the trajectory between the left and right fingers. The motion of fingers is similar to a lever turning around a fixed point. Distances between the endings of finger can represent the range of gripping targets. If y_0 is zero, gripping distance between left and right fingers is shown in Figure 4b. These values of distance can be represented using functions as follows. The variable named y_1 represents distances between the finger's end and the gripper's center line, which can be expressed as:

$$[x_1 + c \times \cos(\theta_1 + \theta_2)]^2 + [y_1 - c \times \sin(\theta_1 + \theta_2)]^2 = c^2, (0 \le x \le x_n) \tag{2}$$

where b is 17.33 mm, a is 62.5 mm, and c is 64.86 mm. Therefore, x_n can be calculated as:

$$x_n = c \times \sin(\theta_1 + \theta_2) - b \tag{3}$$

The variable named y_2 represents distances between the left finger's end and the right's one which can be expressed as:

$$[x_1 + c \times \cos(\theta_1 + \theta_2)]^2 + \left[\frac{y_2}{2} - c \times \sin(\theta_1 + \theta_2)\right]^2 = c^2, \ (0 \leq x \leq x_n)$$

3. Simulation and Calibration of Gripper

The force-displacement and yield limit of the *x-y*-axis adjustable pedestal are analyzed by using finite element analysis (FEA) software. The spring steel with Young's modulus of 200 GPa is selected as the pedestal material. In the simulation, the swivel bolt applies a concentrated force of 50 N on top adjustable block in three strategies. In Figure 5a, the applied force is in negative direction of *y*-axis, and the maximum displacement is obtained as 13.7 μm. Similarly, as the force is applied in positive direction of *x*-axis, the maximum displacement is 13.0 μm, as shown in Figure 5b. In Figure 5c, a maximum displacement of 17.1 μm can be obtained when a combined force along negative *y*-direction and positive *x*-direction is applied on the top block. For these three cases, the displacement of the block is in a uniform distribution. Figure 5d–f shows the corresponding stress distribution of these three cases. It is found that the stress concentration occurs at the flexible hinges. The maximum stress of the hinges is about 262 MPa, which is far less than its yield strength of 520 MPa. It is verified that the proposed design meets the application requirements.

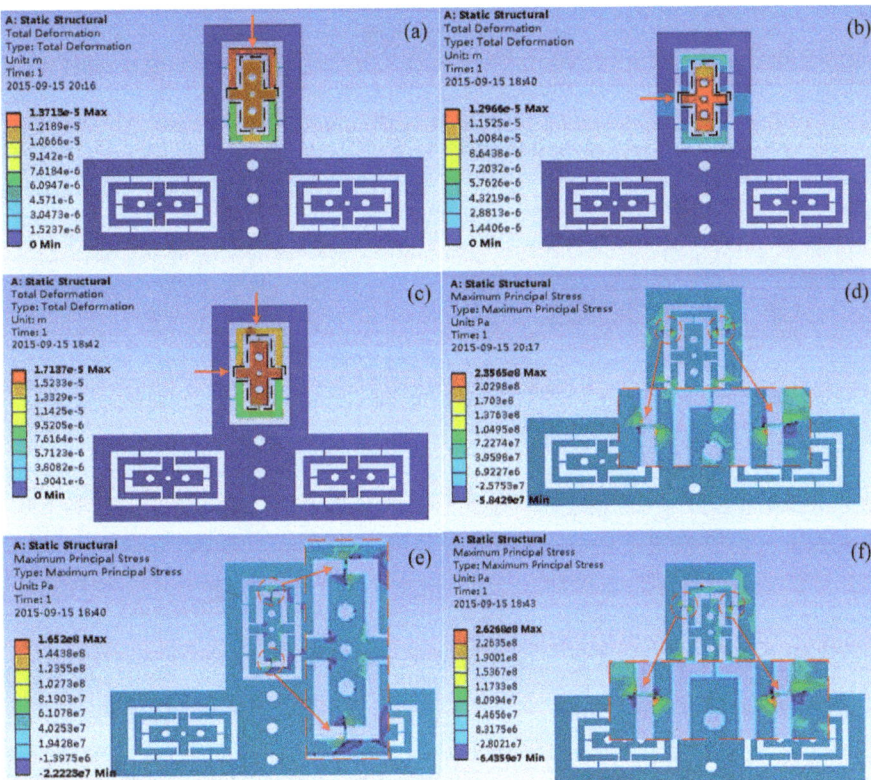

Figure 5. Deformation analysis of the *x-y*-axis adjustable pedestal as the applied force of bolt is in (**a**) *y*-axis, (**b**) *x*-axis, and (**c**) both *x*- and *y*-axis. Stress analysis of the *x-y*-axis adjustable pedestal as the applied force of blot is in (**d**) *y*-axis, (**e**) *x*-axis, and (**f**) both *x*- and *y*-axis.

A similar deformation analysis of the z-axis adjustable pedestal and amplifying structure are conducted and shown in Figure 6. As a concentrated force of 50 N is applied, a displacement of 16.1 μm can be achieved along z-axis as seen in Figure 6a. In Figure 6b, a maximum deformation of 51.3 μm is obtained at the driving force of PZT transducer of 50 N.

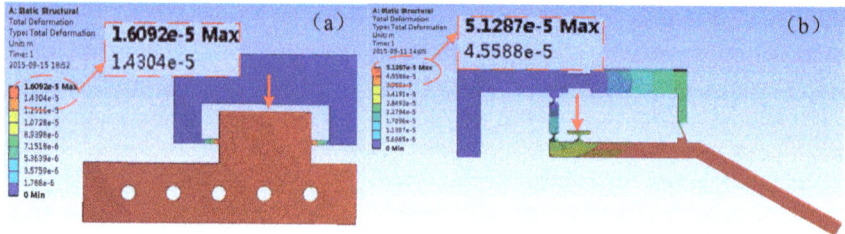

Figure 6. (**a**) Deformation analysis of the z-axis adjustable pedestal as the applied force of bolt is in z-axis. (**b**) Deformation analysis of a single finger as the driving force of PZT transducer is applied.

Figure 7a shows the displacement calibration testing system. It consists of a power driving source, a single finger, a laser sensor, a displacement decoder, and an oscilloscope. When a driving voltage is applied from 0 to 150 V, the PZT transducer can monotonously generate a displacement output from 0 to 10 μm. The end tip displacement of the single finger against driving voltage can be measured by the laser sensor and recorded by the oscilloscope via a displacement decoder. As seen in Figure 7b, the actuation calibration curve is nonlinear, which is due to the arc trajectory of the finger-tip. Besides, errors in machining and assembly process may also have an impact on the output displacement. The result shows that the displacement range of the fingertip is from 0 to 21.7 um as the driving voltage varying from 0 to 150 V. Therefore, the clamping range of the left and right fingers is 43.4 μm. The resolution of the output displacement is 145 nm/V.

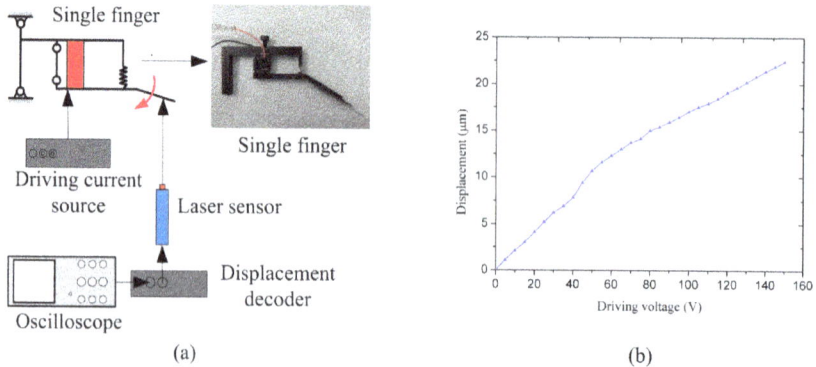

Figure 7. (**a**) Displacement calibration testing system; (**b**) Output displacement of single finger.

4. Experiments and Discussion

As shown in Figure 8, a manipulation system is constructed to demonstrate the capability of the triple-finger gripper. The gripper is installed on a 3-DOF positioning stage (Suruga Seiki, Shisioka, Japan), by which the gripper can be adjusted to the target location. The objects and substrate are placed on a 2-DOF positioning stage, which assist the gripper to locate the objects. A microscope fixed on the other 3-DOF positioning stage helps to identify the objects and get clear images. Three fingers of the gripper are controlled by three piezoelectric ceramics via three channels. A charge-coupled device

(CCD) camera (Daheng Imavision, Beijing, China) is used to record the manipulation processing in real time. Several kinds of end effectors are fixed on the gripper to manipulate various objects.

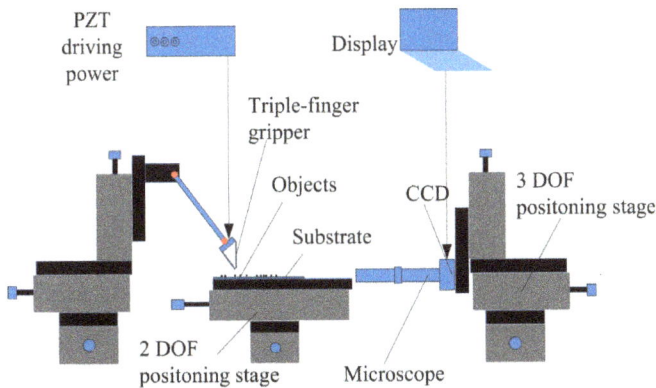

Figure 8. The manipulation setup of the triple-finger gripper.

4.1. Experiment of Double-Finger Gripper

The manipulation experiment of double-finger gripper is firstly conducted. The processes of picking and placing a glass hollow sphere with a diameter of 800 μm and an iron sphere with a diameter of 80 μm are illustrated in Figure 9. Two optical fibers are used as the up and right end effectors of the fingers. Due to the adhesion force, the up and right fingers can pick up the sphere on top of the substrate with a distance of d as shown in Figure 9a. Since the up finger is controlled by the driving PZT transducer, it can provide a vertical vibration to release the adhered sphere as shown in Figure 9b. Similarly, Figure 9c,d show the process of picking and placing an iron sphere with a small diameter of 80 μm. In the experiment, the manipulating objects are different, and the diameter of the objects varies from 80 to 800 μm. It is seen that the gripper is adaptable for multi-target micromanipulation with a quite large clamping range by the adjustable pedestal and PZT driving amplifier structure introducing for each finger.

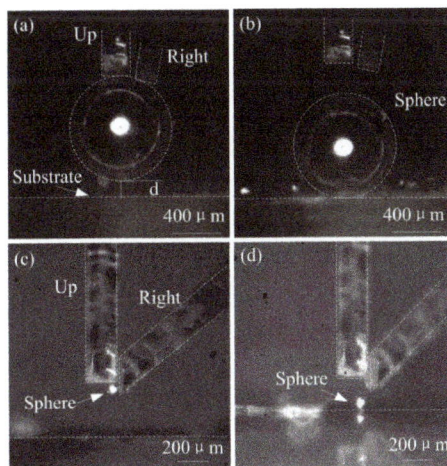

Figure 9. (**a**) Pick and (**b**) place a glass hollow sphere with a diameter of 800 μm; (**c**) Pick and (**d**) place an iron sphere with a diameter of 80 μm.

4.2. Experiment of Triple-Finger Gripper

Figure 10a shows the triple-finger gripper including left, right, and up end effectors on three fingers. The left end effector is an AFM probe. The up and right end effectors are two tungsten tipped probes. The triple-finger gripper is able to clamp boron silicate sphere with a diameter of 10 μm in *y*-direction as seen in Figure 10b. However, without the assist of the third up finger, the sphere failed to release from the fingers due to the adhesion force. It can be seen from Figure 10c that, as the left and right finger open, the sphere is adhered to one of the fingers.

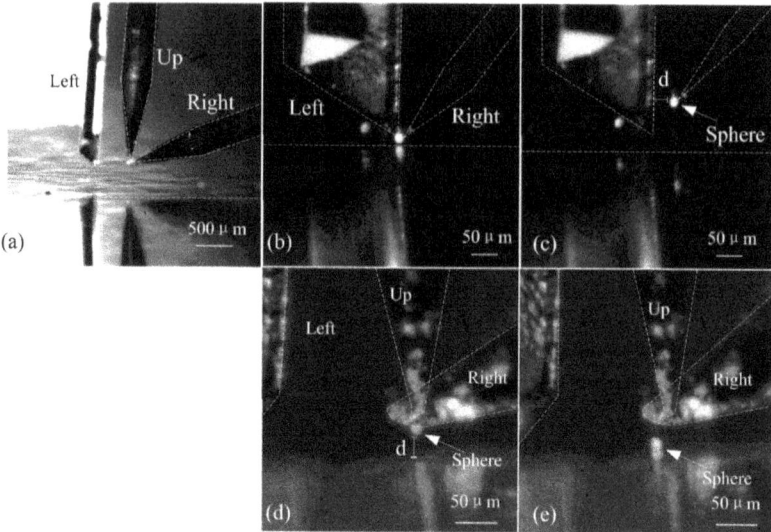

Figure 10. (**a**) Triple-finger gripper; (**b**) Gripping the boron silicate sphere; (**c**) Sphere failed to release from the fingers; (**d**) Fingers with adhered sphere; (**e**) Releasing process with the up finger.

To release the adhered sphere, the up finger is adjusted to the position that above the sphere in Figure 10d. By applying a driving voltage of 20 V and a sinusoidal signal to the PZT transducer, the up finger is able to generate a high frequency vibration, and the sphere is able to release from the right finger easily. Figure 10e shows the enlarged image after releasing process. It is verified that a triple-finger gripper with PZT actuation mechanism gains the advantage to overcome the adhesive force in micro-manipulating process. Therefore, it has a large manipulating range for various micro-objects.

5. Conclusions

This paper demonstrates a PZT actuated gripper which has the ability to pick and release multi-sized objects. The gripper contains three separated fingers, which can be changed in manipulating range through adjustable blocks. The gripper can be fabricated by low-cost machining methods. The three fingers can be assembled with variable end effectors such as tungsten probes, fibers, or AFM probes. The output resolution and displacement range of each single finger is as high as 145 nm/V and 21.7 μm, respectively. To release the objects, the up finger can be controlled to provide vertical vibration. In the experiment, the glass hollow sphere with a diameter of 800 μm and the iron sphere with a diameter of 80 μm were picked and placed by the up and right fingers. The iron sphere of 10 μm in diameter was gripped by the tungsten tipped probes and released by the vibration of the up finger. All these results showed that the triple-finger gripper works well for multi-target micromanipulation.

Future work will focus on force control of the gripping process and optimization of adjustable mechanism. Force control of the griper would improve the efficiency and avoid destruction in manipulation. Optimization of adjustable mechanism would raise the accuracy of gripper and manipulate smaller objects. Additionally, more manipulating strategies will be investigated to pick, transport, and place more targets in the future, which will provide experimental guidance for the design of special micro-manipulation tools. Applications of this devised micro-gripper in super-resolution imaging and inertial confinement fusion adhesion problem solving will be further studied in the future.

Acknowledgments: This work is supported by the National Natural Science Foundation of China (No. 61673287, No. 61433010) and the National High-Tech Research and Development Program of China (No. 2015AA042601).

Author Contributions: All of the authors have equally contributed to this work, either in the implementations or in the writing of this article. T.C., L.S., H.L. managed this work; T.C., Z.Y., L.S. designed the gripper and experiments; Z.Y. and Y.W. analyzed the motion model; Z.Y., H.L., J.L. performed the simulation and calibration of the gripper; Y.W. and J.L. implemented the experiment; T.C., H.L., Y.W., J.L. wrote the paper.

Conflicts of Interest: The authors declare no conflict of interest.

References

1.   Tanikawa, T.; Arai, T. Development of a micro-manipulation system having a two-fingered micro-hand. *IEEE Trans. Robot. Autom.* **1999**, *15*, 152–162. [CrossRef]
2.   Chen, W.; Shi, X.; Chen, W.; Zhang, J. A two degree of freedom micro-gripper with grasping and rotating functions for optical fibers assembling. *Rev. Sci. Instrum.* **2013**, *84*, 115111. [CrossRef] [PubMed]
3.   Sun, Y.; Kim, K. MEMS-based micro and nano grippers with two axis force sensors. U.S. Patent US 8623222, 7 January 2014.
4.   Kim, K.; Liu, X.; Zhang, Y.; Sun, Y. Nanonewton force-controlled manipulation of biological cells using a monolithic MEMS microgripper with two-axis force feedback. *J. Micromech. Microeng.* **2008**, *18*, 055013. [CrossRef]
5.   Xu, Q. Design, fabrication, and testing of an MEMS microgripper with dual-axis force sensor. *Sensors* **2015**, *15*, 6017–6026. [CrossRef]
6.   Boudaoud, M.; Haddab, Y.; Legorrec, Y. Modeling and optimal force control of a nonlinear electrostatic microgripper. *IEEE Trans. Mech.* **2013**, *18*, 1130–1139. [CrossRef]
7.   Chen, B.K.; Zhang, Y.; Perovic, D.D.; Sun, Y. MEMS microgrippers with thin gripping tips. *J. Micromech. Microeng.* **2011**, *21*, 105004–105008. [CrossRef]
8.   Grossard, M.; Boukallel, M.; Régnier, S.; Chaillet, N. Design of integrated flexible structures for micromanipulation. *Flex. Robot.* **2013**. [CrossRef]
9.   Sun, X.; Chen, W.; Tian, Y.; Fatikow, S.; Zhou, R.; Zhang, J.; Mikczinski, M. A novel flexure-based microgripper with double amplification mechanisms for micro/nano manipulation. *Rev. Sci. Instrum.* **2013**, *84*, 085002. [CrossRef] [PubMed]
10. Wang, D.H.; Yang, Q.; Dong, H.M. A monolithic compliant piezoelectric-driven microgripper: Design, modeling, and testing. *IEEE ASME Trans. Mech.* **2013**, *18*, 138–147. [CrossRef]
11. Sun, X.; Chen, W.; Fatikow, S.; Tian, Y.; Zhou, R.; Zhang, J.; Mikczinski, M. A novel piezo-driven microgripper with a large jaw displacement. *Microsyst. Technol.* **2015**, *21*, 931–942. [CrossRef]
12. Liu, J.; Chen, T.; Liu, H.; Chou, X.; Sun, L. PZT driven triple-finger end effectors formicro-manipulation. In Proceedings of the 2015 IEEE International Conference on Cyber Technology in Automation, Control, and Intelligent Systems (CYBER), Shenyang, China, 8–12 June 2015; pp. 1156–1161.
13. Sun, L.; Chen, T.; Chen, L.; Li, X. Design of a hybrid-type electrostatically driven microgripper integrated vacuum tool. *Adv. Mater. Res.* **2009**, *158*, 40–43. [CrossRef]
14. Alogla, A.F.; Amalou, F.; Balmer, C.; Scanlan, P.; Shu, W.; Reuben, R.L. Micro-tweezers: Design, fabrication, simulation and testing of a pneumatically actuated micro-gripper for micromanipulation and microtactile sensing. *Sens. Actuators A Phys.* **2015**, *5*, 394–404. [CrossRef]
15. Chen, T.; Chen, L.; Liu, B.; Wang, J.; Li, X. Design and fabrication of a four-arm-structure MEMS gripper. *IEEE Trans. Ind. Electron.* **2009**, *56*, 996–1004. [CrossRef]

16. Sun, L.; Wang, L.; Rong, W. Capillary interactions between a probe tip and a nanoparticle. *Chin. Phys. Lett.* **2008**, *25*, 1795–1798.

17. Saito, S.; Miyazaki, H.T.; Sato, T.; Takahashi, K. Kinematics of mechanical and adhesional micromanipulation under a scanning electron microscope. *J. Appl. Phys.* **2002**, *92*, 5140–5149. [CrossRef]

18. Ger, T.R.; Huang, H.; Chen, W.; Lai, M. Magnetically-controllable zigzag structures as cell microgripper. *Lab Chip* **2013**, *13*, 2364–2369. [CrossRef] [PubMed]

19. Li, F.; Liu, J. Thermal infrared mapping of the freezing phase change activity of micro liquid droplet. *J. Therm. Anal. Calorim.* **2010**, *102*, 155–162. [CrossRef]

20. Zhang, Y.; Chen, B.; Liu, X.; Sun, Y. Autonomous robotic pick-and-place of microobjects. *IEEE Trans. Robot.* **2010**, *26*, 200–207. [CrossRef]

21. Haliyo, D.S.; Regnier, S.; Guinot, J.C. *[mü]*MAD, the adhesion based dynamic micro-manipulator. *Eur. J. Mech. A Solids* **2003**, *22*, 903–916. [CrossRef]

![micromachines]

micromachines

MDPI

Article

Potential of Piezoelectric MEMS Resonators for Grape Must Fermentation Monitoring [†]

Georg Pfusterschmied [1,*]**, Javier Toledo** [2]**, Martin Kucera** [1]**, Wolfgang Steindl** [1]**, Stefan Zemann** [1]**,
Víctor Ruiz-Díez** [2]**, Michael Schneider** [1]**, Achim Bittner** [1]**,
Jose Luis Sanchez-Rojas** [2]**and Ulrich Schmid** [1]

[1]   Institute of Sensor and Actuator Systems, TU Wien, 1040 Vienna, Austria;
       martin.kucera@tuwien.ac.at (M.K.); steindl.wolfgang@gmail.com (W.S.); stefan.zemann@gmx.at (S.Z.);
       michael.schneider@tuwien.ac.at (M.S.); achim.bittner@tuwien.ac.at (A.B.);
       ulrich.e366.schmid@tuwien.ac.at (U.S.)
[2]   Group of Microsystems, Actuators and Sensors, E.T.S.I. Industriales, Universidad de Castilla-La Mancha,
       13071 Ciudad Real, Spain; javier.toledo.serrano@gmail.com (J.T.); victor.ruiz@uclm.es (V.R.-D.);
       joseluis.saldavero@uclm.es (J.L.S.-R)
*   Correspondence: georg.pfusterschmied@tuwien.ac.at; Tel.: +43-1-58801-36649
†   This paper is an extended version of our paper published in the 27th Micromechanics and Microsystems
       Europe Workshop, Cork, Ireland, 28–30 August 2016.

Received: 27 March 2017; Accepted: 20 June 2017; Published: 26 June 2017

Abstract: In this study grape must fermentation is monitored using a self-actuating/self-sensing piezoelectric micro-electromechanical system (MEMS) resonator. The sensor element is excited in an advanced roof tile-shaped vibration mode, which ensures high Q-factors in liquids (i.e., $Q \sim 100$ in isopropanol), precise resonance frequency analysis, and a fast measurement procedure. Two sets of artificial model solutions are prepared, representing an ordinary and a stuck/sluggish wine fermentation process. The precision and reusability of the sensor are shown using repetitive measurements (10 times), resulting in standard deviations of the measured resonance frequencies of ~0.1%, Q-factor of ~11%, and an electrical conductance peak height of ~12%, respectively. With the applied evaluation procedure, moderate standard deviations of ~1.1% with respect to density values are achieved. Based on these results, the presented sensor concept is capable to distinguish between ordinary and stuck wine fermentation, where the evolution of the wine density associated with the decrease in sugar and the increase in ethanol concentrations during fermentation processes causes a steady increase in the resonance frequency for an ordinary fermentation. Finally, the first test measurements in real grape must are presented, showing a similar trend in the resonance frequency compared to the results of an artificial solutions, thus proving that the presented sensor concept is a reliable and reusable platform for grape must fermentation monitoring.

Keywords: micro-electromechanical system (MEMS); resonator; liquid sensing; piezoelectric; aluminium nitride (AlN); grape must fermentation

---

1. Introduction

The fermentation of grape must into wine involves the interaction between yeasts, bacteria, fungi, and viruses. This complex biochemical process has been recognized and studied since the pioneering investigations of Louis Pasteur in the 1860s. During such a fermentation process, yeasts utilize sugars and other constituents of the grape must for their growth, converting these to alcohol (ethanol), carbon dioxide, and other metabolic end products [1]. A serious problem in winemaking occurs when the yeast growth and the alcoholic fermentation stops prematurely, which results in a wine with residual, unfermented sugar and a concentration of ethanol less than expected. This is indicated in minor deviations of the physical properties, such as density and viscosity, when compared to an ordinary fermentation, and the wine is commonly referred to as being stuck, or sluggish [2,3].

There are several approaches to monitor such a fermentation of wine according to different parameters related to the process. El Haloui et al. [4], Nerantzis et al. [5], and Koukolitschek [6] reported on density determination based on differential pressure measurements or flexural oscillators, respectively. By monitoring the CO_2 release, the fermentation process can also be monitored, as reported in [4]. Another approach is to determine the yeast cell population evolution by means of impedance [7] or turbidity measurements [8]. The propagation velocity in grape must, determined by ultrasound measurements, can also be used for monitoring the fermentation process [9,10]. Lastly, a reflective technique based on fiber optics was reported in [11,12]. Basically, all of these different approaches have their specific drawbacks; some lack enough accuracy, others have only been tested in discrete must samples and, in some cases, the sensor output performance deteriorates dramatically due to an increasing deposition of tartaric acid crystals on the active surface of the sensors [12]. In recent years, cantilever-like micro-electromechanical system (MEMS) resonators have become a reliable platform for various sensing applications [13]. One significant advantage of such resonators is that they can usually be fabricated using silicon micromachining technology and are, therefore, smaller and much cheaper to produce, compared to traditional quartz crystal resonators. So-called "fluidic channel resonators", in which the fluid passes through a fluidic channel in the moving part of the resonator, is beneficial due to their high-precision capabilities, especially for molecular detection [14], blood coagulation [15], or to quantify other physical properties of fluids, such as mass density or viscosity [16,17]. With such a system only a small amount of liquid (5 μL [18]) is needed and masses can be detected down to the attogram regime [19]. For actuation and sensing, lasers are often used to drive such micro- [20] or nano- [21] mechanical resonators. This approach is considered to be a very accurate and flexible, with the drawback of making the measurement setup bulky, expensive, and mobile integration impracticable. A disadvantage of fluidic channel resonators is their low reusability, which is caused by potential clogging of the fluidic channel. Therefore, such resonators are often designed as one-time use devices, which increases the cost of the overall system. In contrast, solid resonators are often easier to clean, which increases reusability, but generally lacks in precision compared to fluidic channel resonators. The recent improvements in the field of piezoelectric solid MEMS resonators for liquid monitoring purposes predestines such a system for sensing applications, where decent accuracy is required [22–24]. The piezoelectric actuation and readout mechanism keeps the sensor device reasonably small and does not require any further laser-based measurement equipment. In this particular field, the question arises whether a micro-machined solid resonator is capable to detect the minor deviations in the physical properties of the grape must during the fermentation process, and whether it offers a promising alternative to the presented measurement approaches. Most recently, Toledo, et al. [25] introduced a phase locked loop-based oscillator circuit in combination with a commercial lock-in amplifier to track the oscillation frequency of the solid resonators [26]. Using this measurement setup, in combination with an evaluation procedure presented in [27], it could be shown that the monitoring of grape must fermentation using solid MEMS resonators is, in principal, possible.

In this study we will focus on the reliability and the reusability of such a solid resonator and what impact potential contaminations caused have on the sensor characteristics despite the harsh liquid environment. Based on these investigations, an estimation of the precision of the sensor concept is given. Furthermore, different measurement and evaluation procedures are used to validate the data presented in [25]. Finally, test measurements in a real grape must are presented and compared to those of the artificial solutions, thus proving the suitability of the presented concept for grape must fermentation monitoring.

2. Experimental Details

2.1. Sensor Specification

The piezoelectric MEMS resonator is fabricated on four-inch SOI-wafers and features a length of $L = 2524$ µm, a width of $W = 1274$ µm, and a thickness of $T = 20$ µm, passivated with a SiO_2/Si_3N_4 bi-layer of $t_{iso} = 250/80$ nm. For actuating and sensing, an aluminium nitride (AlN) thin film with a thickness of $t_{AlN} = 1$ µm is sputter-deposited onto the plate surface and is sandwiched between two chromium/gold thin film electrodes with equal thickness for the bottom and top electrode of $t_{be} = t_{te} = 50/450$ nm. In Figure 1, a typical sensor chip after packaging and wire bonding with a released single side-clamped resonator, and its top view as an inset, are shown.

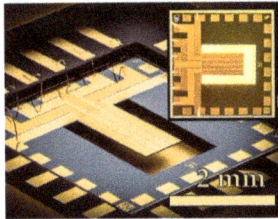

Figure 1. Optical micrograph of the in-house fabricated silicon die (6 mm × 6 mm), containing one piezoelectric actuated plate (dimensions: 2524 µm × 1274 µm × 20 µm) using advanced electrode patterning considering the volume-strain of the modal shape presented in Figure 2.

The corresponding mode shape is illustrated by finite-element method (FEM) eigenmode analyses of the fourth-order (15-mode) of the roof tile-shaped mode in Figure 2a in side and, in Figure 2b, in top view. In Figure 3 a schematic cross-sectional view is presented, showing the electrode design optimized for the fourth-order of the roof tile-shaped mode and four electrodes ($w_o = 335$ µm, $w_i = 272$ µm) with alternating anti-parallel (↓↑↓↑) electric excitation to ensure in addition a collection of almost all generated polarization charges without cancellation. The result of this tailored electrode design is an increased deflection in z-direction and, in further consequence, an increased strain related conductance peak as reported in [28], which is one of the major reasons why this oscillation mode is used in this study. Furthermore, it is worth mentioning that the resonance frequencies in liquids do not exceed 600 kHz, which simplifies the electrical measurement procedure and minimized any pressure-induced energy losses by sound waves, which requests compressive fluid properties as reported in [29]. Secondly, high Q-factors above 100 are achieved with this type of mode, which facilitates a high sensor sensitivity. To overcome the obstacle of parasitic current caused by the conductivity of the model solutions, the entire sensor element, including bond wires and ceramic package, is passivated with an amorphous silicon dioxide thin film with a thickness of ~4 µm.

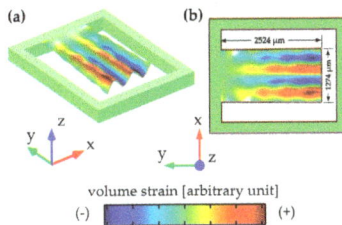

Figure 2. Visualization in side view in (**a**) and in top view in (**b**) for a plate excited in the fourth-order of the roof tile-shaped mode (15-mode). The colored areas on the cantilever surface represent the local volume strain distribution.

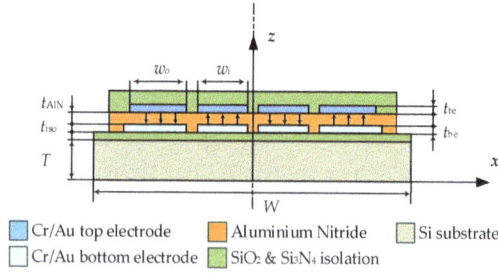

Figure 3. Schematic cross-sectional view on the micro-electromechanical system (MEMS) resonator illustrating the electrode design and the plate support of the 15-mode.

2.2. Artificial Model Solutions and Sensing Principle

Two sets of model solutions are prepared, using a R200D microscale from Sartorius (Goettingen, Germany), representing an ordinary (N_{1-9}) and a stuck/sluggish (S_{1-9}) wine fermentation process, consisting of fructose, glucose, glycerol, and ethanol [30]. The specific compositions are listed in Table 2 and Table 3 and show significant differences in the fructose, glucose, and ethanol concentrations, when comparing ordinary (Table 2) and stuck/sluggish (Table 3) wine fermentation processes. The resonator is immersed in these model solutions and actuated in the fourth-order of the roof tile-shaped mode with an Agilent 4294 A impedance analyzer (excitation voltage $V_{exc} = 500$ mV AC, Keysight, Böblingen, Germany). The actuation in the resonance leads to an increased average deflection and to increased strain on the sensor surface, respectively. Due to the piezoelectric effect, polarization charges are generated, which are detected by the impedance analyzer as an increased conductance peak ΔG. A typical peak characteristic of the piezoelectric MEMS resonator is shown in Figure 4 when immersed in the model solution N_1. This measurement procedure lasts only ~10 s. Subsequently, the sensor is cleaned with dish soap and isopropanol, followed by a 20 min drying process at room temperature to avoid tartaric acid crystal deposition, as reported in [12].

From these output characteristics, the resonance frequency f_{res} and Q-factor Q are determined using the Butterworth-van Dyke equivalent circuit, in combination with a Levenberg-Marquardt algorithm [31]. Once f_{res} and Q are obtained for all solutions (N_1–N_9 and S_1–S_9), an evaluation procedure taken from [32] is used to determine the actual density ρ_f using:

$$\rho_f = \frac{\text{Im}(Z_m) - \omega_{res}L_m + \frac{1}{\omega_{res}C_m} - \text{Re}(Z_m) + R_m}{\omega_{res}d_1} \tag{1}$$

Here, Z_m describes the impedance of the equivalent resonance circuit including a resistance R_m, an inductance L_m and a capacity C_m at a certain angular frequency ω_{res} in liquid. The constant parameter d_1 is determined in a calibration liquid with known viscosity and density, using:

$$d_1 = \frac{\text{Im}(Z_m) - \omega_{res}L_{m\_air} + \frac{1}{\omega_{res}C_{m\_air}} - d_2\sqrt{\mu_f\rho_f\omega_{res}}}{\omega_{res}\rho_f}, \tag{2}$$

with:

$$d_2 = \frac{\text{Re}(Z_m) - R_{m\_air}}{\sqrt{\mu_f\rho_f\omega_{res}}} \tag{3}$$

where R_{m_air}, L_{m_air}, and C_{m_air} are the corresponding unperturbed resonator values in air and are given in Table 1. For this initial calibration process values supplied by the manufacturer for μ_f and ρ_f of N_4 and S_4 are used to calculate d_1 and d_2, as shown in Equations (2) and (3), respectively. Once these two constants are obtained, all further values of ρ_f can be determined using Equation (1).

Table 1. Characterization of the resonator in air. R_{m_air}, L_{m_air}, and C_{m_air} represent the unperturbed resonator in air.

f_{res}	Q	R_{m_air}	L_{m_air}	C_{m_air}
[kHz]	[1]	[kΩ]	[mH]	[fF]
1014.1	329.6	1.365	70.662	348.77

2.3. Real Grape Must Fermentation

For the investigation of a real fermentation process, *Airen* grapes are processed into grape must with a subsequent filtering process to avoid damaging the sensor by large grape skin particles. After this procedure, the grape must is inoculated with 0.2 g/mL of *Saccharomyces cerevisiae* strain (UCLM S325, Fould-Springer, Ciudad Real, Spain) previously rehydrated, as described in the supplier guidelines. A cylindrical fermenter (mini-Bioreactor Applikon Buitechnology B.V., Delft, The Netherlands) is filled with 3 L of inoculated must, where the temperature was controlled at 28 °C. The fermentation monitoring is carried out by means of the extraction and analysis of 7 mL-samples approximately every 5 h during the following six days. After extraction, the samples were centrifuged for one minute at 1000 rpm (Universal 32R Hettich, Tuttlingen, Germany) and kept refrigerated until analysis.

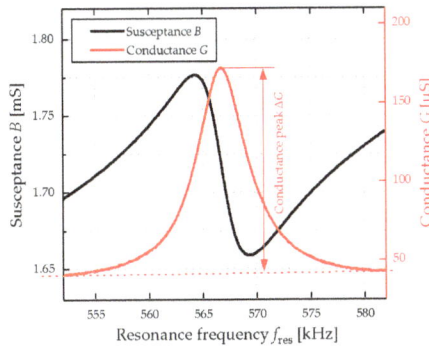

Figure 4. Representative electrical output characteristics given the conductance G and susceptance B of piezoelectric MEMS resonators when excited in the fourth-order of the roof tile-shaped mode in model solution N_1.

Table 2. Composition of the model solutions representing a desired wine fermentation process from N_1 (raw grape must) up to N_9 (ordinary fermented model solution) [30]. The measured values are obtained using a R200D microscale from Sartorius.

Solution	Fructose		Glucose		Glycerol		Ethanol	
-	[g/L]		[g/L]		[g/L]		[% *v/v*]	
-	Nom.	Meas.	Nom.	Meas.	Nom.	Meas.	Nom.	Meas.
N_1	110	109.90	100	99.90	0	0	0	-
N_2	90	89.90	80	80.00	0	0	1	-
N_3	70	69.95	30	30.03	5	4.99	6	-
N_4	60	59.97	20	19.99	5	5.00	8	-
N_5	40	39.90	10	9.95	6	5.99	9	-
N_6	20	19.90	2	2.00	7	6.97	12	-
N_7	8	8.02	2	2.00	7	6.99	13	-
N_8	5	5.00	2	2.00	7	6.99	13	-
N_9	2	2.00	1	0.99	9	8.99	14	-

Table 3. Composition of the model solutions representing a stuck wine fermentation process from S_1 (raw grape must) up to S_9 (stuck fermented model solution) [30]. The measured values are obtained using a R200D microscale from Sartorius.

Solution	Fructose		Glucose		Glycerol		Ethanol	
-	[g/L]		[g/L]		[g/L]		[% v/v]	
-	Nom.	Meas.	Nom.	Meas.	Nom.	Meas.	Nom.	Meas.
S_1	110	109.90	100	99.98	0	0	0	-
S_2	90	90.00	80	79.98	0	0	1	-
S_3	80	79.98	60	60.01	3	3.00	3	-
S_4	70	69.98	30	29.98	5	4.99	6	-
S_5	66	65.99	27	27.03	5	4.99	7	-
S_6	64	64.00	26	26.06	5	5.01	7	-
S_7	62	61.99	24	24.00	6	5.99	7	-
S_8	58	57.99	22	22.00	6	5.99	7	-
S_9	57	56.99	21	21.00	7	7.01	8	-

3. Results

In Figure 5, the results from the reproducibility measurements are depicted for N_3 and N_7 of an ordinary fermented grape must in Figure 5a, and for S_7 of a stuck fermented grape must in Figure 5b. The measurement procedure, including the cleaning process presented in the previous part of this paper, is repeated for 10 times. Thereby, standard deviations in the resonance frequency f_{res} of 0.102% for N_3, 0.085% for S_7 and 0.094% for N_7 are obtained, thus showing a high potential for the targeted application. Higher standard deviations are obtained for the Q-factor (~11%) and the conductance peak (12%), which, however, have minor impacts on the final calculation of the density values, indicated by a standard deviation of the calculated density values of 1.2% for N_3, 1.1% for N_7, and 1.0% for S_7.

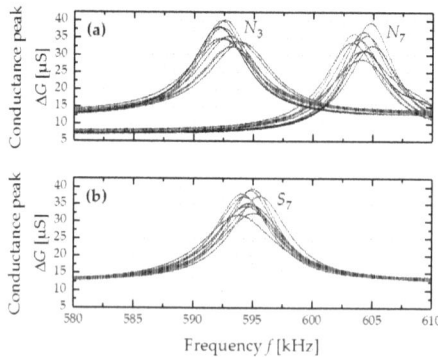

Figure 5. Precision evaluation on the basis of two model solutions of an ordinary fermented grape must in (**a**), and of a stuck fermented grape must in (**b**) at 22 °C.

The compositions of the artificial wine solutions for an ordinary (N_{1-9}) wine fermentation process, listed in Table 2, shows a decrease in fructose and glucose concentration from 110 and 100 g/L (N_1) to 2 and 1 g/L (N_9), respectively. Likewise, the glycerol and ethanol concentration are increasing from zero to 9 g/L and 14% v/v. In the case of a stuck fermented wine process, as it is listed in Table 3, the fructose and glucose concentrations decrease to 50 and 20 g/L, respectively, and remain constant at these values. In parallel, the increase in the ethanol concentration stops as well, and does not exceed a value of 8% (v/v). These significant changes in the composition of the model solutions affect the resonance response of the MEMS sensor as shown for both an ordinary fermentation in Figure 6 and a stuck fermentation in Figure 7. In the first case, the resonance frequency increases monotonically for the ordinary fermentation (see Figure 6). In contrast, the increase of the resonance frequency during the investigation of the stuck fermentation plateaus at ~575 kHz.

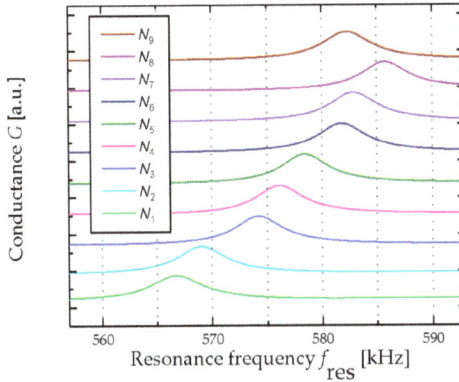

Figure 6. Electrical output characteristics of the piezoelectric MEMS resonator for an ordinary fermentation process N_1–N_9, starting with N_1 at the bottom and all other model solutions (N_2–N_9) stacked above. The *y*-axis is scaled as arbitrary unit [a.u.].

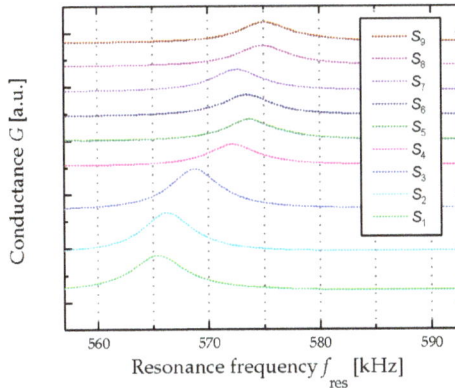

Figure 7. Electrical output characteristics of the piezoelectric MEMS resonator for a stuck fermentation process S_1–S_9, starting with S_1 at the bottom and all other model solutions (S_2–S_9) stacked above. The *y*-axis is scaled as arbitrary unit [a.u.].

In Figure 8, the results of the resonance response analysis are compared to the ethanol concentration $\sigma_{Ethanol}$ for both sets of model solutions. Significant differences in the resonance frequencies f_{res} are evident with increased $\sigma_{Ethanol}$ values when comparing ordinary and stuck fermentation, as well as a good correlation between $\sigma_{Ethanol}$ and f_{res}. The combination of unfermented sugars and lower concentration of ethanol in a stuck fermentation (see Table 3) leads to a flat resonance characteristics, allowing to detect stuck fermentation. The results from the repeatability measurements of N_3, N_7, and S_7 from Figure 5 are used to estimate the precision of the MEMS sensor. For a further analysis of this particular progress a second resonator with the same geometry is excited in the first roof tile-shaped mode (12-mode). The resonance frequencies in both sets of model solutions amount around 1/4 of the resonance frequency of the 15-mode (~160 kHz) and are depicted in Figure 9, showing a similar characteristics in the frequency response. Compared to the results presented in Figure 7 the absolute change in resonance frequency is also ca. four times lower, leading to a lower sensitivity for the 12-mode. Therefore, all further evaluations are performed using the results from the resonator excited in the 15-mode. In Figure 10 the resonance frequencies for both parameter sets (N_{1-9} and S_{1-9}) as a function of the corresponding density values are depicted, showing a linear increase in f_{res} from 567 kHz (N_1) to 582 kHz (N_9), and again a premature stop at ~575 kHz for the stuck fermentation.

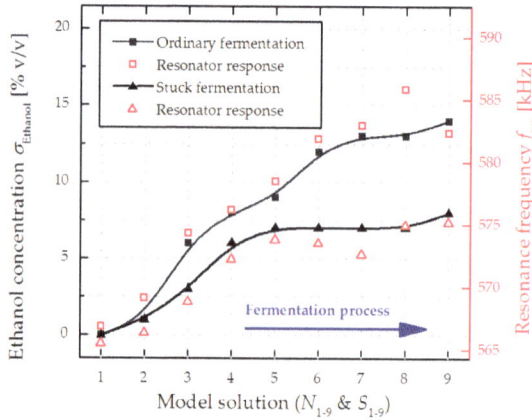

Figure 8. Fifteen-mode frequency response analysis of an ordinary (N_{1-9}) and a stuck (S_{1-9}) fermentation process in comparison to the nominal ethanol concentration of the investigated model solutions. The inserted lines serve as guide to the eye.

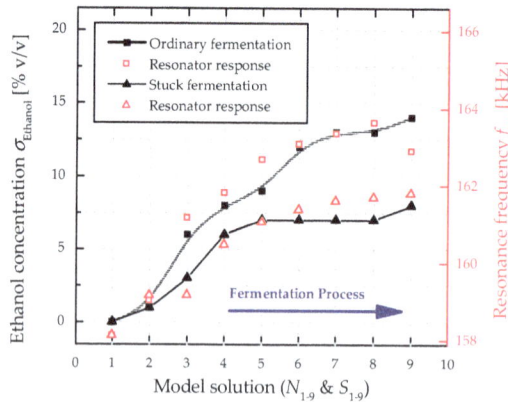

Figure 9. Twelve-mode frequency response analysis of an ordinary (N_{1-9}) and a stuck (S_{1-9}) fermentation process in comparison to the nominal ethanol concentration of the investigated model solutions. The inserted lines serve as guide for the eye.

As a next step, the density values of the model solutions are evaluated with a Stabinger SVM3000 viscometer (Anton Paar, Graz, Austria). The results are listed in Table 4 and depicted in Figure 10, showing an almost linear decrease for N_{1-9} from 1.081 (N_1) to 0.996 g/cm^3 (N_9) and a premature stop in the density decrease for S_{1-9} with a minimal value of 1.020 g/cm^3 (S_7). Furthermore, the results of the density determination are compared to those from the Stabinger viscometer in Figure 11, showing low deviations and the possibility to distinguish between ordinary and stuck fermentations at an early stage of the fermentation process.

Table 4. Determined density values (Res.) for ordinary and stuck/sluggish model solutions and their comparison to reference values (Stabinger) obtained with a Stabinger SVM3000 viscometer.

	Ordinary				Stuck/Sluggish		
Sol.	Density ρ [g/cm^3]			Sol.	Density ρ [g/cm^3]		
-	Stabinger	Res.	Dev.	-	Stabinger	Res.	Dev.
N_1	1.081	1.082	−0.001	S_1	1.075	1.075	-
N_2	1.065	1.063	0.002	S_2	1.063	1.064	−0.001
N_3	1.032	1.033	−0.001	S_3	1.050	1.050	-
N_4	1.020	1.023	−0.003	S_4	1.031	1.031	-
N_5	1.007	1.010	−0.003	S_5	1.028	1.023	0.005
N_6	0.994	0.994	-	S_6	1.025	1.024	0.001
N_7	0.988	0.989	−0.001	S_7	1.020	1.029	−0.009
N_8	0.985	0.978	0.007	S_8	1.023	1.020	0.003
N_9	0.996	0.995	0.001	S_9	1.021	1.020	0.001

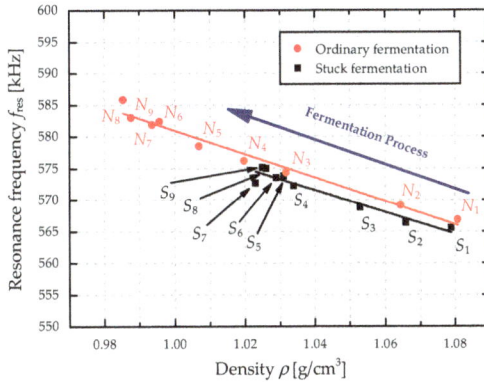

Figure 10. Correlation between the measured resonance frequencies and the corresponding density values for an ordinary (**red**) and a stuck (**black**) fermentation process.

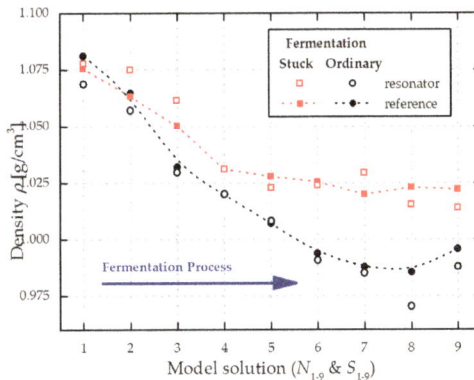

Figure 11. Density values for ordinary ($N_{1–9}$) and stuck ($S_{1–9}$) fermented model solutions evaluated with the piezoelectric MEMS resonator (unfilled dots) and its reference values obtained with a Stabinger SVM3000 viscometer (filled dots). The inserted dashed lines serve as guides for the eye.

Finally, the results of the frequency response analysis for real grape must fermentation are presented in Figure 12, showing a similar trend in the change of f_{res}, compared to those of the artificial solutions presented in Figure 8, which indicates the high potential of the presented sensor system to withstand the harsh conditions during the fermentation process even in real grape must.

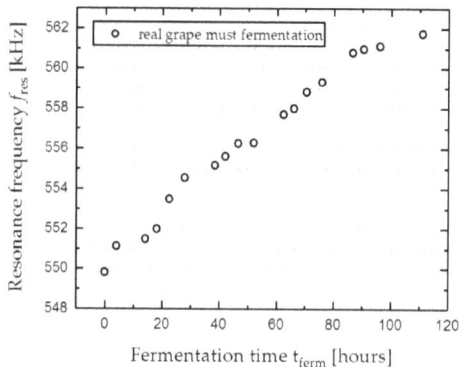

Figure 12. Fifteen-mode frequency response in a real grape over fermentation time at 20 °C.

4. Conclusions

In this paper, an approach to monitor the fermentation processes in winemaking by analyzing the frequency response of a piezoelectric MEMS resonator is presented. The sensor is excited at the fourth-order of the roof tile-shaped mode in several artificial grape must model solutions, representing an ordinary and a stuck wine fermentation process, respectively. During the artificial fermentation, both an increasing ethanol and a decreasing total sugar concentration reduce the density of the grape must despite the counteracting effect of the increasing glycerin concentration. This decrease in density leads to higher resonance frequencies. Due to the high Q-factor of the MEMS sensor in liquid media (~100 in isopropanol), even minor shifts in resonance frequency can be detected with high precision, which enables the possibility to monitor these changes of the physical properties during the fermentation processes. Reproducibility and reusability measurement were performed, showing low standard deviations in the resonance frequencies of ~0.1% and moderate deviations in the Q-factor (~11%) and conductance peak (~12%). These results show that the resonance frequency is hardly effected by potential surface contaminations. Due to the fast measurement cycle of 10 s and easy cleaning procedure, harsh contaminations, such as the deposition of tartaric acid crystals, could be avoided. Nevertheless, minor surface contaminations reduce the precision of the calculated density values leading to a standard deviation of ~1.1%. Finally, test measurements in real grape must were performed resulting in similar sensor characteristics compared to the results of the model solutions, demonstrating the high potential of the presented sensor concept for grape must fermentation monitoring.

Acknowledgments: This work has been supported by the Austrian Research Promotion Agency within the COMET-K2 Project XTribology (project No. 849109) and by the Spanish Ministerio de Economía y Competitividad project TEC2015-67470-P. Víctor Ruiz-Díez and Javier Toledo acknowledge financial support from the Spanish Ministry of Education, Culture, and Sport (grants FPU-AP2010-6059 and FPI-BES-2013-063743). Furthermore, Sherif Soliman and Mert-Ziya Erses are gratefully acknowledged.

Author Contributions: The concept and design of the presented sensor is contributed to Martin Kucera, Achim Bittner, Jose Luis Sanchez-Rojas and Ulrich Schmid. The fabrication of the MEMS sensor was carried out by Georg Pfusterschmied and Wolfgang Steindl. The evaluation in artificial and real grape must samples was performed by Georg Pfusterschmied, Stefan Zemann and Javier Toledo, respectively. Georg Pfusterschmied analyzed the data and wrote the paper. Víctor Ruiz-Díez, Jose Luis Sanchez-Rojas, Michael Schneider and Ulrich Schmid provided guidance and supervision.

Conflicts of Interest: The authors declare no conflict of interest.

References

1. Fleet, G.H. *Wine Microbiology and Biotechnology*; CRC Press: Boca Raton, FL, USA, 1993.
2. Larue, F.; Lafon-Lafourcade, S. Survival factors in wine fermentation. In *Alcohol Toxicity in Yeasts and Bacteria*; CRC Press: Boca Raton, FL, USA, 1989; pp. 193–215.
3. Munoz, E.; Ingledew, W.M. Yeast hulls in wine fermentations—A review. *J. Wine Res.* **1990**, *1*, 197–209. [CrossRef]
4. El Haloui, N.; Picque, D.; Corrieu, G. Alcoholic fermentation in winemaking: On-line measurement of density and carbon dioxide evolution. *J. Food Eng.* **1988**, *8*, 17–30. [CrossRef]
5. Nerantzis, E.; Tataridis, P.; Sianoudis, I.; Ziani, X.; Tegou, E. Winemaking process engineering on line fermentation monitoring—Sensors and equipment. *Sci. Technol.* **2007**, *5*, 29–36.
6. Koukolitschek, K. Verfahren und Vorrichtung zur Präzisen Bestimmung der Alkoholkonzentration in Flüssigkeiten. German Patent DE413,841,9C2, 2 September 1993.
7. Pérez, M.A.; Muñiz, R.; De La Torre, C.; García, B.; Carleos, C.E.; Crespo, R.; Cárcel, L.M. Impedance spectrometry for monitoring alcoholic fermentation kinetics under wine-making industrial conditions. In Proceedings of the XIX IMEKO World Congress Fundamental and Applied Metrology, Lisbon, Portugal, 6–11 September 2009; pp. 2574–2578.
8. Crespo, R.; Cárcel, L.; Pérez, M.; Nevares, I.; Del Álamo, M. Suitable at-line turbidity sensor for wine fermentation supervision. In Proceedings of the International Conference on Food Innovation 2010, Valencia, Spain, 25–29 October 2010; pp. 1–4.
9. Lamberti, N.; Ardia, L.; Albanese, D.; Di Matteo, M. An ultrasound technique for monitoring the alcoholic wine fermentation. *Ultrasonics* **2009**, *49*, 94–97. [CrossRef] [PubMed]
10. Resa, P.; Elvira, L.; De Espinosa, F.M.; González, R.; Barcenilla, J. On-line ultrasonic velocity monitoring of alcoholic fermentation kinetics. *Bioprocess Biosyst. Eng.* **2009**, *32*, 321–331. [CrossRef] [PubMed]
11. Acevedo, J.M.; Gandoy, J.D.; Del Río Vázquez, A.; Freire, C.M.-P.; Soria, M.L. Plastic optical fiber sensor for real time density measurements in wine fermentation. In Proceedings of the IEEE Instrumentation and Measurement Technology Conference (IMTC), Warsaw, Poland, 1–3 May 2007.
12. Graña, C.Q.; Acevedo, J.M. Experiences in measuring density by fiber optic sensors in the grape juice fermentation process. In Proceedings of the XIX IMEKO World Congress Fundamental Applied Metrology, Lisbon, Portugal, 6–11 September 2009; pp. 2579–2582.
13. Boisen, A.; Dohn, S.; Keller, S.S.; Schmid, S.; Tenje, M. Cantilever-like micromechanical sensors. *Rep. Prog. Phys.* **2011**, *74*, 036101. [CrossRef]
14. Burg, T.P.; Manalis, S.R. Suspended microchannel resonators for biomolecular detection. *Appl. Phys. Lett.* **2003**, *83*, 2698–2700. [CrossRef]
15. Cakmak, O.; Ermek, E.; Kilinc, N.; Bulut, S.; Baris, I.; Kavakli, I.H.; Yaralioglu, G.G.; Urey, H. A cartridge based sensor array platform for multiple coagulation measurements from plasma. *Lab Chip* **2015**, *15*, 113–120. [CrossRef] [PubMed]
16. Godin, M.; Bryan, A.K.; Burg, T.P.; Babcock, K.; Manalis, S.R. Measuring the mass, density, and size of particles and cells using a suspended microchannel resonator. *Appl. Phys. Lett.* **2007**, *91*, 123121. [CrossRef]
17. Khan, M.F.; Schmid, S.; Larsen, P.E.; Davis, Z.J.; Yan, W.; Stenby, E.H.; Boisen, A. Online measurement of mass density and viscosity of pL fluid samples with suspended microchannel resonator. *Sens. Actuators B Chem.* **2013**, *185*, 456–461. [CrossRef]
18. Bircher, B.A.; Duempelmann, L.; Renggli, K.; Lang, H.P.; Gerber, C.; Bruns, N.; Braun, T. Real-time viscosity and mass density sensors requiring microliter sample volume based on nanomechanical resonators. *Anal. Chem.* **2013**, *85*, 8676–8683. [CrossRef] [PubMed]
19. Lee, J.; Shen, W.; Payer, K.; Burg, T.P.; Manalis, S.R. Toward attogram mass measurements in solution with suspended nanochannel resonators. *Nano Lett.* **2010**, *10*, 2537–2542. [CrossRef] [PubMed]
20. Cakmak, O.; Ermek, E.; Kilinc, N.; Yaralioglu, G.G.; Urey, H. Precision density and viscosity measurement using two cantilevers with different widths. *Sens. Actuators A Phys.* **2015**, *232*, 141–147. [CrossRef]
21. Bircher, B.A.; Krenger, R.; Braun, T. Automated high-throughput viscosity and density sensor using nanomechanical resonators. *Sens. Actuators B Chem.* **2016**, *223*, 784–790. [CrossRef]

22. Kucera, M.; Wistrela, E.; Pfusterschmied, G.; Ruiz-Díez, V.; Manzaneque, T.; Sánchez-Rojas, J.L.; Schalko, J.; Bittner, A.; Schmid, U. Characterization of a roof tile-shaped out-of-plane vibrational mode in aluminum-nitride-actuated self-sensing micro-resonators for liquid monitoring purposes. *Appl. Phys. Lett.* **2014**, *104*, 233501. [CrossRef]

23. Kucera, M.; Wistrela, E.; Pfusterschmied, G.; Ruiz-Díez, V.; Sánchez-Rojas, J.L.; Schalko, J.; Bittner, A.; Schmid, U. Characterisation of multi roof tile-shaped out-of-plane vibrational modes in aluminium-nitride-actuated self-sensing micro-resonators in liquid media. *Appl. Phys. Lett.* **2015**, *107*, 053506. [CrossRef]

24. Pfusterschmied, G.; Kucera, M.; Wistrela, E.; Steindl, W.; Ruiz-Díez, V.; Bittner, A.; Sánchez-Rojas, J.L.; Schmid, U. Piezoelectric response optimization of multi roof tile-shaped modes in MEMS resonators by variation of the support boundary conditions. In Proceedings of the 2015 Transducers—2015 18th International Conference on Solid-State Sensors, Actuators and Microsystems (TRANSDUCERS), Anchorage, AK, USA, 21–25 June 2015; pp. 969–972.

25. Toledo, J.; Jiménez-Márquez, F.; Úbeda, J.; Ruiz-Díez, V.; Pfusterschmied, G.; Schmid, U.; Sánchez-Rojas, J.L. Piezoelectric MEMS resonators for monitoring grape must fermentation. *J. Phys. Conf. Ser.* **2016**. [CrossRef]

26. Toledo, J.; Manzaneque, T.; Ruiz-Díez, V.; Jiménez-Márquez, F.; Kucera, M.; Pfusterschmied, G.; Wistrela, E.; Schmid, U.; Sánchez-Rojas, J.L. Out-of-plane piezoelectric microresonator and oscillator circuit for monitoring engine oil contamination with diesel. In Proceedings of the Smart Sensors, Actuators, and MEMS VII, and Cyber Physical Systems, Barcelona, Spain, 4 May 2015.

27. Toledo, J.; Manzaneque, T.; Ruiz-Díez, V.; Jiménez-Márquez, F.; Kucera, M.; Pfusterschmied, G.; Wistrela, E.; Schmid, U.; Sánchez-Rojas, J.L. Comparison of in-plane and out-of-plane piezoelectric microresonators for real-time monitoring of engine oil contamination with diesel. *Microsyst. Technol.* **2016**, *22*, 1–10. [CrossRef]

28. Pfusterschmied, G.; Kucera, M.; Ruiz-Díez, V.; Bittner, A.; Sánchez-Rojas, J.L.; Schmid, U. Multi roof tile-shaped vibration modes in MEMS cantilever sensors for liquid monitoring purposes. In Proceedings of the 28th IEEE International Conference on Micro Electro Mechanical Systems (MEMS), Estoril, Portugal, 18–22 January 2015.

29. Van Eysden, C.A.; Sader, J.E. Frequency response of cantilever beams immersed in compressible fluids with applications to the atomic force microscope. *J. Appl. Phys.* **2009**, *106*, 094904. [CrossRef]

30. Jiménez-Márquez, F.; Vázquez, J.; Úbeda, J.; Sánchez-Rojas, J.L. High-resolution low-cost optoelectronic instrument for supervising grape must fermentation. *Microsyst. Technol.* **2014**, *20*, 769–782. [CrossRef]

31. Manzaneque, T.; Hernando, J.; Rodriguez-Aragon, L.; Ababneh, A.; Seidel, H.; Schmid, U.; Sánchez-Rojas, J.L. Analysis of the quality factor of AlN-actuated micro-resonators in air and liquid. *Microsyst. Technol.* **2010**, *16*, 837–845. [CrossRef]

32. Pfusterschmied, G.; Kucera, M.; Wistrela, E.; Manzaneque, T.; Ruiz-Díez, V.; Sánchez-Rojas, J.L.; Bittner, A.; Schmid, U. Temperature dependent performance of piezoelectric MEMS resonators for viscosity and density determination of liquids. *J. Micromech. Microeng.* **2015**, *25*, 105014. [CrossRef]

micromachines

MDPI

Article

MEMS Gyroscopes Based on Acoustic Sagnac Effect [†]

Yuanyuan Yu [1], Hao Luo [2], Buyun Chen [1], Jin Tao [1], Zhihong Feng [1], Hao Zhang [1], Wenlan Guo [1] and Daihua Zhang [1,*]

[1] State Key Laboratory of Precision Measurement Technology and Instruments, School of Precision
    Instruments and Opto-Electronics Engineering, Tianjin University, Tianjin 300072, China;
    yuanyuanyu@tju.edu.cn (Y.Y.); buyunc@tju.edu.cn (B.C.); taojin@tju.edu.cn (J.T.); zhfeng@tju.edu.cn (Z.F.);
    haozhang@tju.edu.cn (H.Z.); guowenlan@tju.edu.cn (W.G.)
[2] Intel Labs, San Francisco, CA 95054, USA; memsluo@gmail.com
[*] Correspondence: dhzhang@tju.edu.cn; Tel.: +86-22-2740-7565
[†] This paper is an extended version of our paper published in the 14th IEEE SENSORS Conference,
    Busan, Korea, 1–4 November 2015.

Academic Editors: Ulrich Schmid, Michael Schneider and Nam-Trung Nguyen
Received: 31 October 2016; Accepted: 19 December 2016; Published: 24 December 2016

Abstract: This paper reports on the design, fabrication and preliminary test results of a novel
microelectromechanical systems (MEMS) device—the acoustic gyroscope. The unique operating
mechanism is based on the "acoustic version" of the Sagnac effect in fiber-optic gyros. The device
measures the phase difference between two sound waves traveling in opposite directions, and
correlates the signal to the angular velocity of the hosting frame. As sound travels significantly
slower than light and develops a larger phase change within the same path length, the acoustic gyro
can potentially outperform fiber-optic gyros in sensitivity and form factor. It also promises superior
stability compared to vibratory MEMS gyros as the design contains no moving parts and is largely
insensitive to mechanical stress or temperature. We have carried out systematic simulations and
experiments, and developed a series of processes and design rules to implement the device.

Keywords: acoustic gyroscope; Sagnac effect; phase difference; sound waves; angular velocity;
sensitivity

---

1. Introduction

The gyroscope (gyro) is an inertial sensor determining the speed of rotational motions [1]. It has
a wide range of applications including the fields of consumer electronics, automotive, aerospace and
navigation [2–4]. Recently, gyros have witnessed a new wave of market growth driven by increasing
needs of Internet of Things (IoT) and wearable devices. Their functions have been extended from
basic motion sensing and navigation to much wider areas including human-machine interaction,
health monitoring, and fitness analysis [5–7]. Among them, microelectromechanical systems (MEMS)
vibratory gyros and fiber optic gyros are the most prevalent platforms in portable systems.

MEMS vibratory gyros are particularly attractive because of their small size, low cost, low
power consumption, and easy integration with complementary metal-oxide-semiconductor (CMOS)
circuitry [8–10]. They are based on the energy transfer between two orthogonal vibration modes as
a result of the Coriolis effect. However, typical MEMS gyros come with a number of intrinsic drawbacks.
The vibratory structure makes the device highly susceptible to external shock and vibration [11,12].
The structural design is inevitably associated with parasitic capacitive coupling and quadrature errors,
which collectively deteriorate the performance and limit wider adoption of the MEMS gyros.

Fiber-optic gyros, on the other hand, are based upon a different sensing mechanism (the Sagnac
effect), and do not involve the above issues. The device correlates angular speeds with interference
signals from two laser beams. It usually consists of thousands of fiber coils to maximize the optical

path length in order to achieve low drift and a high scale factor [13,14]. However, this generally makes the gyro too big and expensive for most consumer applications [12].

To overcome the drawbacks of the two classes of gyros, we propose a new MEMS device—the acoustic gyro—in this work. It is based on the "acoustic version" of the Sagnac effect in the fiber-optic gyro. The device measures the phase difference between two sound waves traveling in opposite directions in a circular air duct, and correlates the phase difference with the rotational speed of the waveguide. Since sound travels much slower than light, the acoustic gyro can develop a much larger phase difference within the same path length compared to fiber-optic gyros, or, equivalently, it can produce comparable sensing signals within a much shorter path length and significantly reduce the device size. In addition, the fabrication of the acoustic gyro is fully compatible with CMOS processes and it is readily integrated with peripheral electronics. In contrast to conventional MEMS gyros, the acoustic gyro is intrinsically immune to mechanical stress and temperature changes. The device also has a very simple structure that substantially lowers the manufacturing complexity.

This report expands on our previous work published in [15]. In this paper, we will present both the theoretical evaluations and the very recent progress in process development and testing. We have systematically evaluated and optimized a series of design rules to achieve acceptable process stability, device yield and performance. According to our theoretical predictions and preliminary testing results, the device can be packed into a 5 mm × 5 mm footprint and can potentially achieve a high sensitivity.

2. Materials and Methods

2.1. Working Principle of Novel Device

Figure 1 depicts the idea of the acoustic Sagnac effect in detail. A 3D finite element model (FEM) using COMSOL multiphysics software (v5.0, COMSOL Co., Ltd., Stockholm, Sweden) is set up to study the principle theoretically (Figure 1a). This model consists of a circular air duct, a sound transmitter, and a set of receivers. It couples the acoustic-piezoelectric interaction module and aeroacoustic module together. The former is used to simulate the sound transmitter and receivers with sound hard boundary condition, including sound generation and electro-acoustical conversion efficiency. The latter is adopted to emulate the pressure wave propagation in the air duct with no-slip boundary condition of air flow. The simulation results show the principle as below. When the transmitter at 12 o'clock position generates a pressure pulse, it induces two sound waves propagating in the clockwise and counterclockwise directions, respectively. When the transmitter, the duct and the air in it are stationary, the two wave fronts travel at the same speed and always meet at 6 and 12 o'clock positions (top in Figure 1a). On the other hand, if the duct and the transmitter move relative to the air inside or vice versa, the two sound waves propagate at different speeds (bottom in Figure 1a) and develop a phase difference with each other over time. Placing sound receivers along the air duct would allow us to quantify the phase difference and correlate the results with the rotating speed of the frame. The phase difference related with frame angular velocity can be detected by the subsequent digital phase detector. According to Reference [15], the phase shift $\Delta\varphi$ and sensitivity S_φ (defined as the phase shift per unit rotational velocity) of this device are

$$\Delta\varphi = 2\pi f \cdot \Delta t = \frac{8\pi^2 R^2 \Omega}{v^2} \cdot f \tag{1}$$

$$S_\varphi = \Delta\varphi/\Omega = \frac{8\pi^2 R^2}{v^2} \cdot f \tag{2}$$

where R, f, v and Ω are the radius of the air duct, acoustic frequency, acoustic velocity and the rotating speed of the frame, respectively. We can see from this formula the phase sensitivity is inversely proportional to the square of propagation velocity. The phase sensitivity between the acoustic and fiber-optic gyros can be calculated as

$$\frac{\Delta\varphi_{acoustic}}{\Delta\varphi_{optic}} = \frac{f_{acoustic}}{f_{optic}} \cdot \frac{c_{optic}^2}{c_{acoustic}^2} = \frac{10[\text{MHZ}]}{193[\text{THz}]} \cdot \frac{(3 \times 10^8 [\text{m/s}])^2}{(343[\text{m/s}])^2} \approx 4 \times 10^4 \tag{3}$$

where f_{optic} is the light frequency using 193 THz [16–18], and $f_{acoustic}$ is the sound frequency as 10 MHz which can propagated in air [19,20]. In this calculation we assumed the same ring radius R and angular velocity Ω for both devices. The estimation indicated that the acoustic gyro could theoretically achieve a phase sensitivity of 10^4 higher than that of an optical device with the same dimension. In other words, the acoustic device could deliver comparable performance with significantly smaller size.

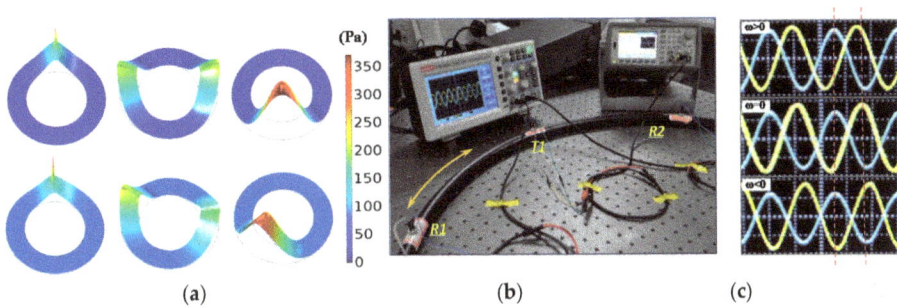

Figure 1. (**a**) Simulated sound wave propagating in a circular air duct when the air is stationary (above) and flowing (below). The setting of (below) is equivalent to the situation where the air keeps stationary while the frame (together with the sound transmitter and receiver(s)) rotates in the opposite direction. The out-of-plane displacement is the height expression of acoustic pressure (Pa); (**b**) Emulation test with a hula-hoop as the circular air duct. Diameter of the duct is about 80 cm. Three 40 kHz commercial ultrasonic ceramic transducers are installed inside the duct as acoustic transmitter and receivers. The transmitter and two receivers are marked as T1, R1 and R2, respectively; (**c**) Test results of (**b**). The phase difference between R1 and R2 changes when the ring rotates in clockwise and counterclockwise directions at $\approx 10°/\text{s}$. The oscilloscope time base is 10 µs/div.

The Sagnac effect in optics derives from the assumption that the speed of light is independent of the motion of its source or medium. Apparently the same rule does not apply to ultrasound. In fact, the Acoustic Sagnac effect results from the inertia of air—the transmission medium keeps stationary when the circular duct rotates. Therefore, the two acoustic waves travel at the same speed in the inertial reference frame of the observer. This effect has been further verified by the emulation test in Figure 1c,d, in which a hula-hoop was used as the air duct. Three 12.6-mm-diameter 40 kHz commercial ultrasonic ceramic transducers (TCT40-12T/R-1.2) are installed inside the hula-hoop duct working as an acoustic transmitter and two receivers, respectively. The two ultrasound receivers located at different positions (labeled as R1 and R2) were able to detect different phase shifts when the duct was rotated at various speeds. The test result shows the time difference between the two acoustic waves is about 1.5 µs at an angular velocity of ~20°/s. It is in good agreement with the theoretical prediction.

2.2. Design and Structure

Figure 2 illustrates our scheme to build the gyro on a miniature MEMS chip. Figure 2a,b show the top and side views of the device with dimensions. The device is fabricated by bonding a cap wafer (containing a circular trench and through silicon vias (TSVs)) to a base wafer hosting a set of aluminum nitride piezoelectric micromachined ultrasonic transducers (AlN-PMUTs). The AlN-PMUTs are used as the acoustic transmitter as well as the receivers. The trench on the cap wafer defines the circular air duct to guide the acoustic waves. We selected PMUTs over capacitive micromachined ultrasonic transducers (CMUTs) due to the fact that the former does not need exceedingly high voltage bias (usually hundreds of volts for CMUT) or ultrafine microstructures to achieve sufficient transducer sensitivity, which effectively reduces circuit and fabrication complexity and cost [21–23].

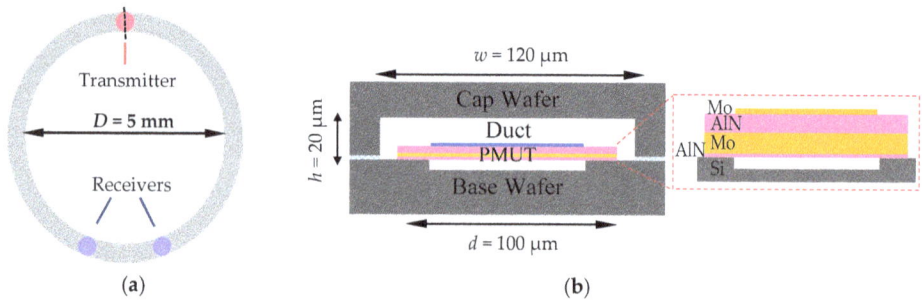

Figure 2. (**a**) Top-view schematic of the acoustic gyro (not in scale). In this specific case, three aluminum nitride piezoelectric micromachined ultrasonic transducers (AlN-PMUTs) are fabricated on the base wafer, one as a transmitter and two as receivers. The two receivers are placed symmetrically around the vertical axis. The circular trench on the cap wafer, once bonded to the base wafer, forms an enclosed air duct and bridges the three PMUTs; (**b**) Cross-sectional view across the center of the transmitter as marked by the dashed line in (**a**). The inset shows cross-section view of PMUT.

Before assigning specific dimensions to each component, the operating frequency f_0 needs to be determined first. It is related to several factors. According to Equation (1), higher frequency yields higher sensitivity of phase shift to rotational speed. However, it may also lead to larger acoustic attenuation, making the ultrasound waves hard to be detected at the receivers. In addition, one needs to consider the relationship between the wavelength ($\lambda = v/f_0$) and the PMUTs' diameter (*d*) as well. First, *d* needs to be smaller than λ to avoid near-field irregular pressure pattern [24]. Second, λ is proportional to the square of *d* as a result of flexural-mode resonance. Therefore, *d* needs to be greater than a lower limit to satisfy both requirements. This imposes an upper limit to f_0. Taking into account all the requirements and limitations, we finally set the operating frequency f_0 to be approximately 1.6 MHz. The diameter of PMUT should not be too small as compared to the wavelength for efficient transmission [25,26], or greater than the wavelength resulting in irregular near-field pressure pattern [24]. An ideal choice for the diameter of PMUTs in this work is 100 μm [25].

Given the target frequency (f_0 = 1.6 MHz) and PMUT diameter, we are now able to determine the stack thickness of the PMUTs (f_0 scales linearly with the film thickness and $1/d^2$ [25]) based on the given material. However, the layer thickness will impact other performance parameters such as electromechanical coupling coefficient, transmitting sensitivity (Pa/V) or receiving sensitivity (V/Pa). In this work, each PMUT on the base wafer consists of 4 functional layers suspended on an air cavity, which are the top electrode (TE, made of Mo), piezoelectric layer (PZ, made of AlN), bottom electrode (BE, made of Mo), and the seed layer (SL, made of AlN) from top to bottom. A big advantage of AlN is that its relatively low dielectric constant minimizes device capacitance, thereby producing a higher voltage between the top and bottom surfaces when operated as an ultrasonic receiver. The BE and SL layers serve as a passive layer to induce a vertical stress gradient across the whole stack, which forces the suspended membrane to deflect vertically when a transverse stress originates in the PZ layer (due to the piezoelectric coefficient e$_{31,f}$). Different BE and SL thicknesses (with respect to the PZ layer) result in different transmitting and receiving sensitivities of the PMUTs. In our case, the net efficiency of the entire ultrasound generating, propagating, and receiving processes can be written as:

$$S = \frac{v_{output}}{v_{input}} = \frac{P_{tx}}{v_{input}} \cdot \frac{v_{output}}{P_{rx}} \cdot \frac{P_{rx}}{P_{tx}} = G_T \cdot G_R \cdot G_{ch} \tag{4}$$

where v_{input} and v_{output} are the voltages applied to and detected at the transmitter and the receiver(s), respectively. P_{tx} and P_{rx} are the magnitude of the pressure wave right above the two PMUTs. The three parameters G_T, G_R and G_{ch} characterize the transmitting sensitivity (in Pa/V), receiving sensitivity (in V/Pa) and the acoustic transmission attenuation, respectively. Previous studies [27–30] have observed non-monotonic dependence of G_T and G_R on the thickness of the passive layer. Qualitatively speaking, a thicker passive layer facilitates flexural bending of the entire membrane, but decreases the overall electromechanical coupling coefficient k_{eff}^2 at the same time [31–33].

We have setup a finite element model to quantitatively determine the optimum thickness. Figure 3a plots the surface pressure (in Pa) of the PMUT under a 1 V_{pp} driving voltage at resonant frequency as a function of PZ/BE thickness ratio. Peak position of each curve indicates the optimum thickness ratio for maximum G_T. Figure 3b evaluates the receiving sensitivity for different PZ/BE stacks and plots the electrical potential developed under a constant pressure difference across the PMUT (100 Pa). The peaks correspond to maximum G_R values. Thickness of the SL is fixed at 50 nm in both figures. In Figure 3c,d, we repeat the same analysis with fixed BE thickness (100 nm) and varying PZ/SL ratios. According to the simulation results, multiply of the two coefficients ($G_T \cdot G_R$) reaches its maximum at 500 (PZ)/500 (BE)/50 (SL) nm with the desired resonant frequency (1.6 MHz). This stack setting is used in our final design to optimize the net efficiency (S) of the transmitter-receiver pair.

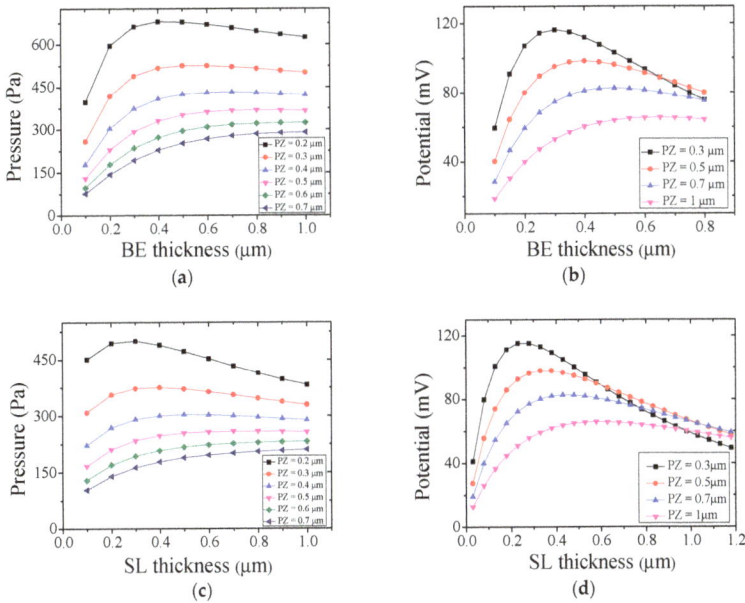

Figure 3. (a) Surface pressure under 1 V_{pp} driving voltage at the resonant frequency; (b) Electrical potential under 100 Pa as a function of piezoelectric layer/bottom electrode (PZ/BE) thickness ratio, with the seed layer (SL) thickness fixed at 50 nm; (c,d) The surface pressure and the electrical potential under various PZ/SL thickness ratios when the BE thickness is fixed at 100 nm.

It is worth noting that the top electrode (TE) of the PMUT has a smaller diameter compared to the rest layers in order to optimize the electromechanical coupling coefficient. We set the TE diameter as 70 μm, 70% of the PMUT diameter based upon the design rules discussed in previous reports [20,34,35].

We then use the following rules to set the geometric dimensions of the circular air duct. The channel width (w) should be shorter than the acoustic wavelength λ to avoid large sidelobes, and large enough to enclose individual PMUTs inside the channel ($w > d$). The perimeter of the circular duct (πD, D being the diameter) should be integer times of $\lambda/2$ to facilitate standing-wave formation. In addition, internal height (h) of the air duct needs to be sufficiently shorter than λ to minimize energy dissipation at boundaries. Based on these considerations, we set $w = 130$ μm, $D = 5$ mm, and $h = 20$ μm in our final design.

Acoustic-piezoelectric frequency domain simulations are then carried out to verify the functionality of all components. In the model, we apply a continuous driving voltage of $V_{input} = 10$ V across the transmitter PMUT (at $f_0 = 1.6$ MHz) and visualize the formation and detection of the acoustic waves. Figure 4a plots the distribution of the air pressure field, confirming the generation of a standing ultrasonic wave along the circular duct when the air duct is stationary. The inset zooms into the regions near the transmitter and the receivers to display finer features. Figure 4b examines the detection of the ultrasound when the air duct is stationary. It allows us to calculate the surface potential (V_{output}) on the receiver PMUTs, which is around 65 mV in this specific setting. This corresponds to a net efficiency (S) of ~0.06 according to the definition in Equation (4). The simulation result demonstrates good detectability of the ultrasonic waves by the receivers, and proves good feasibility of our structural design.

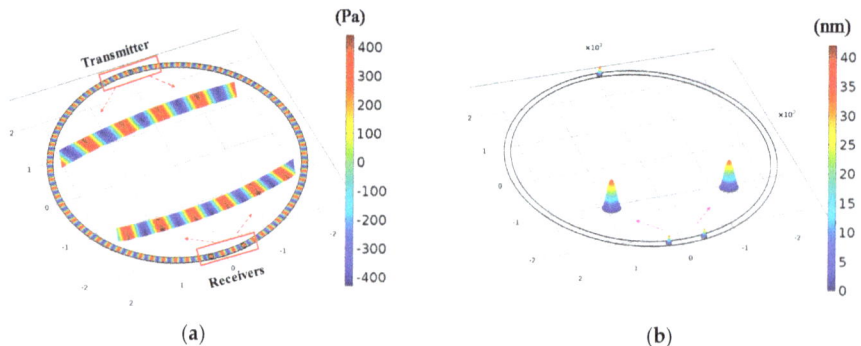

(a)   (b)

Figure 4. Finite element model (FEM) simulation results. (**a**) Acoustic pressure field distribution inside the air duct ($f_0 = 1.6$ MHz). Color bar: Acoustic pressure (Pa); (**b**) Maximum mechanical displacement of the transmitter and receivers. Color bar: Displacement (nm).

In addition to the above design, we have incorporated other design variations on the mask to maximize wafer usage. They include three kinds of ducts as circular, spiral and square. The circular ducts are designed within eight variants using different duct width, different PMUT diameter and quantities on duct. Two spiral ducts are presented by varying duct width and PMUTs diameter to improve device performance by maximizing path length. Each spiral shape includes four spiral structures which can eliminate cross sensitivity and the common–mode output errors by differential operation. The square ducts with different widths and PMUT diameters are used to evaluate possible dependence on the shape of the acoustic path. Figure 5 shows a section of the mask layout. The entire mask set contains 9 layers including all the PMUT, air duct, and TSV structures.

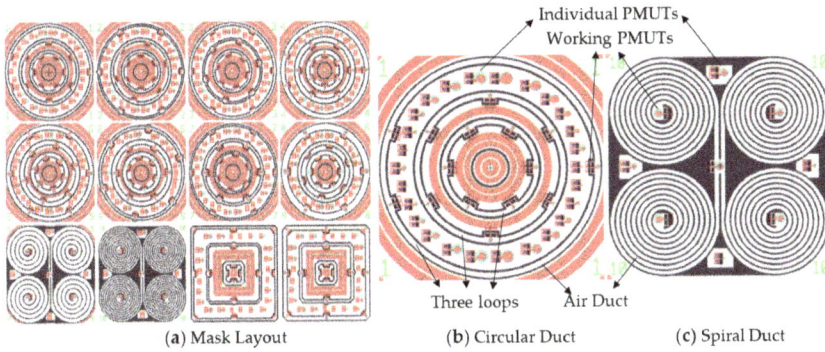

(a) Mask Layout   (b) Circular Duct   (c) Spiral Duct

Figure 5. Mask layout with circular, spiral and square shaped air ducts. (**a**) Mask Layout; (**b**) Circular Duct and (**c**) Spiral Duct.

2.3. Fabrication

Figure 6 summarizes the entire process flow. The fabrication involves three major stages. The first stage deals with the fabrication of PMUTs on the base wafer (Figure 6a–f). Specifically, it starts with a sacrificial release pit formed by dry etching (Figure 6a) and CVD deposition of phosphosilicate glass (PSG) followed by chemical mechanical planarization (CMP) (Figure 6b). A 50 nm AlN seed layer and 500 nm Mo BE) are then deposited by sputtering at elevated temperature. The AlN seed layer provides a high-quality <110> crystalline foundation for the BE, PZ, and TE layers built on its top. It also electrically insulates the metal electrodes from the underlying Si substrate. The BE layer is then patterned by dry etch (Figure 6c). This is followed by deposition of 500 nm AlN and 150 nm Mo to form the PZ and TE layers, which are then patterned (through wet and dry etch to define the contours and expose metal pads and release holes (Figure 6d)). We use E-beam evaporation and lift-off to deposit Cr (0.2 μm)/Au (1 μm) pads in selected areas as the interface layer for wafer bonding (Figure 6e). In the final step, HF solution is used to remove the sacrificial layer and suspend the film (Figure 6f).

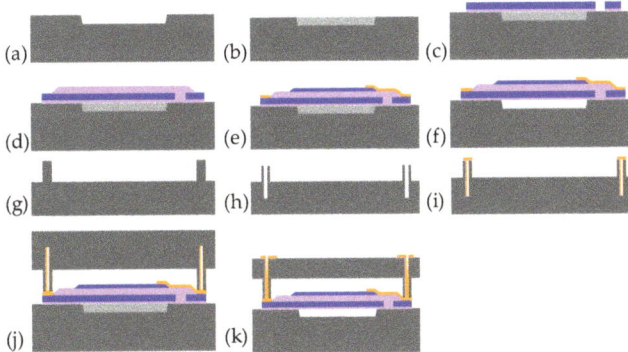

Figure 6. Base wafer flow: (**a**) Silicon etching; (**b**) SiO$_2$ deposition and chemical mechanical planarization (CMP); (**c**) BE deposition and pattern; (**d**) AlN/TE deposition and pattern; (**e**) Au deposition and pattern; (**f**) sacrificial layer release; Cap wafer flow: (**g**) trench etch and pattern; (**h**) vias etch (**i**) 0.2 μm Cr/1 μm Au deposition and pattern; Bonding part: (**j**) Au-Au bonding; (**k**) CMP, electroplate, and Au deposition.

Figure 6g–i depicts the process flow on the cap wafer. We first use dry etch to create a circular trench with side walls (Figure 6g), then make high-aspect-ratio holes through deep RIE (reactive ion etch) at the locations of TSVs (through silicon vias) (Figure 6h). The edges and inner walls of these holes are then covered with Cr (0.2 μm)/Au (1 μm) that works as the wafer bonding adhesive and the electrical interconnects between the base and the cap wafer (Figure 6i).

In the third stage, we flip over the cap wafer and bond it against the base wafer (Figure 6j). The bonding creates an enclosed air duct bridging all PMUTs on the same ring. Au-Au adhesion provides good bonding strength and establishes electrical connects between the two wafers at the same time. It is important to note that the bonding chamber is filled with N_2 and maintained at an inner pressure of 1 atm throughout the process. The cap wafer also protects the PMUTs from damages by back-end processes including wafer dicing and plastic molding (when needed). The whole wafer is then thinned to 500 μm by mechanical grinding to expose the pre-defined TSVs. The vias are then filled with Cu by electroplating. This is followed by deposition and patterning of metal pads (Ti/Au) on the top surface to complete the entire flow (Figure 6k).

The flow has proven to be a highly stable process and yields good uniformity and consistency. Figure 7a shows the microscope images of whole base wafer, circular device structure and individual PMUTs on it. Figure 7b–e are the scanning electron microscope (SEM) images taken from perspective angles to present close-up views of different structures.

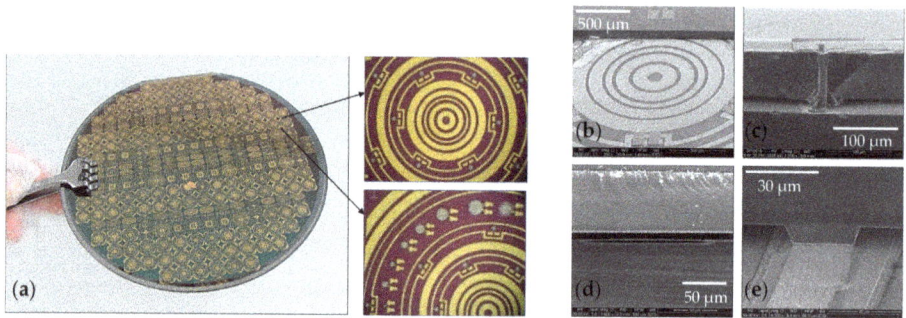

Figure 7. (**a**) Microscope images of whole wafer, circular device structure and individual PMUTs; (**b**) scanning electron microscope (SEM) image of a device with partially peeled-off cap; (**c**) close-up view of a TSV; (**d**) cross-sectional view of a PMUT; and (**e**) close-up view of a bonding interface.

3. Results

We first evaluate the transmitting performance of the PMUTs. The surface displacement as a function of the driving frequency is measured in the air by a Laser Dropper Vibrometer (LDV, OFV 512 and OFV 2700, Polytec, Inc., Waldbronn, Germany) as shown in Figure 8. The PMUTs are excited with a 1 V sinusoidal signal. The measured resonant frequency is 1.616 MHz, very close to the designed value of 1.6 MHz. The maximum displacement at the resonant frequency is 221 nm. The measurements on four PMUTs (Figure 8a) at different locations indicate good frequency uniformity across the wafer. This ensures good frequency matching between the acoustic transmitter and receiver and maximizes the net efficiency S. The surface deflection (d_f) is related to the local air pressure by

$$P = (2\pi f d_f)\rho_0 c_0 A_e \tag{5}$$

where ρ_0 is the density of air, c_0 is the sound velocity, and $A_e = 1/3$ is a correction factor accounting for the deviation from an ideal piston model [36,37]. The transmitting sensitivity G_T is then calculated to be ~330 Pa/V, slightly lower compared to the simulation result in Figure 3a, presumably due to geometrical mismatches and additional loss mechanisms through air damping, anchors and boundaries.

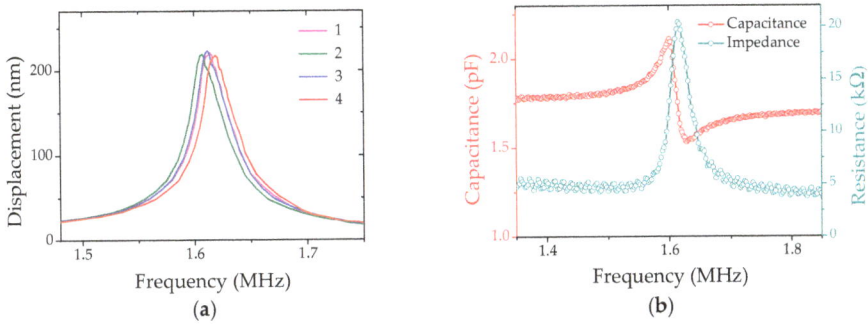

Figure 8. (a) Frequency dependence of surface displacements of four 100 μm PMUTs. The resonance frequencies are narrowly distributed with a very small variation of ~0.7%; (b) Impedance measurement of a single 100 um PMUT with a very high coupling coefficient of $k_t^2 = 3.64\%$.

Impedance measurements of the PMUTs (100 μm in diameter) have been carried out using an impedance analyzer, the Agilent4294A (Agilent Technologies, Inc., Palo Alto, CA, USA, Figure 8b). The result shows good agreement on the resonant frequency with the LDV measurement. In addition, the impedance analysis provides both the resonance (f_r) and anti-resonance (f_a) frequencies, and allows us to calculate the electromechanical coupling coefficient k_{eff}^2 according to [23,38]

$$\frac{k_{eff}^2}{1 - k_{eff}^2} = \frac{f_a^2 - f_r^2}{f_r^2} \tag{6}$$

Given $f_r = 1.606$ MHz and $f_a = 1.636$ MHz from the data in Figure 8b, k_{eff}^2 is calculated to be 3.64%. This value is significantly higher than typical numbers in previous reports on AlN-based PMUTs (0.056% in [23] and 0.387% in [39]). High material and process quality are likely acceptable for the improvement, as is using the mass production PVD tool for the film deposition. In addition, the device design also plays a role in this variance. The BE (Mo) layer serves as the passive layer with a high Young's modulus. This increases the stiffness of the passive layer, thus raising the electromechanical coupling coefficient [40,41]. Moreover, the low parasitic capacitance in this PMUT contributes to the higher coupling factor as well.

Next, we measure the receiving sensitivity by setting up two PMUTs facing each other. The two devices are wire-bonded to separate PCB boards, which are mounted on translational stages to continuously adjust the separation. The left PMUT is driven by a continuous sine wave (10 V_{pp}) with a frequency swept from 1.3 to 2 MHz. The right is used as a receiver to pick up the ultrasound signal and feed it through a lock-in amplifier. Outputs of the amplifier, with both the amplitude and phase signals, are shown in Figure 9a. The distance between the two PMUTs is 1.36 mm. The maximum amplitude is about 52.5 μV at the resonant frequency of 1.617 MHz. The signal diminishes when we place thin sheets of dielectric materials (e.g., paper or glass) in front of the receiver, which confirms that the signal originates from the ultrasonic waves rather than the electromagnetic couplings. Furthermore, when we replace the dielectric sheets with a grounded metal mesh to block possible electromagnetic interference, the resonance signal persists. In Figure 9b, we record the receiver signals at different distances. The decay with increasing distance is in good agreement with acoustic attenuation in air.

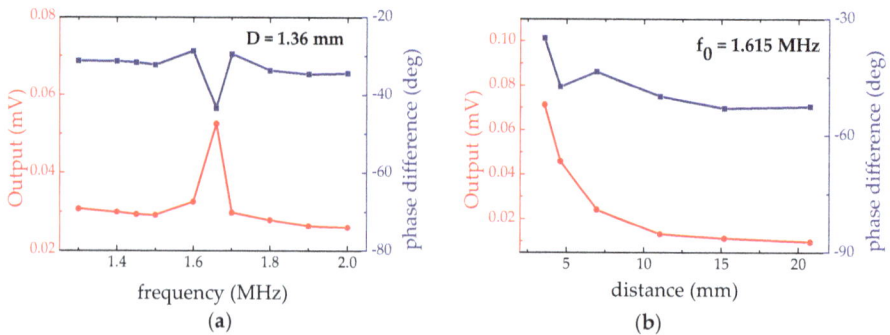

Figure 9. (**a**) Readout of the receiver at 1.36 mm from the transmitter; (**b**) Receiver signals taken at different distances from the source.

4. Conclusions

In this paper, we present the design, fabrication and preliminary testing results of a new MEMS gyro. It measures angular velocity by comparing the phase difference between two acoustic waves. The device promises a similar performance to fiber-optic gyros with significant downsizing by four to five orders of magnitude. We have proven the concept by systematic FEM simulations, and successfully completed the fabrication and preliminary evaluation of the structural design. Further tests are ongoing to provide more comprehensive results and understanding of the design and sensing mechanism.

Acknowledgments: This work was partly supported by Tianjin Applied Basic Research and Advanced Technology (13JCYBJC37100) and the Program of Introducing Talents of Discipline to Universities (111 project No. B07014).

Author Contributions: Y.Y., H.L. and D.Z. conceived and designed the experiments; Y.Y., B.C. and J.T. performed the experiments. H.F., H.Z. and W.G. provided advice on experiment design and data analysis. D.Z. analyzed the data and wrote the paper. All authors have read and approved the final manuscript.

Conflicts of Interest: The authors declare no conflict of interest.

References

1. Kaajakari, V. Gyroscope. In *Practical MEMs*; Small Gear Publishing: Las Vegas, NV, USA, 2009; pp. 346–364.
2. Trusov, A.A.; Schofield, A.R.; Shkel, A.M. Performance characterization of a new temperature-robust gain-bandwidth improved MEMS gyroscope operated in air. *Sens. Actuators A Phys.* **2009**, *155*, 16–22. [CrossRef]
3. Johari, H. Micromachined Capacitive Silicon Bulk Acoustic Wave Gyroscopes. Ph.D. Thesis, Georgia Tech School of Mechanical Engineering, Atlanta, GA, USA, 2008.
4. Nasiri, S. *A Critical Review of MEMS Gyroscopes Technology and Commercialization Status*; Invensense: Sunnyvale, CA, USA, 2008; p. 8.
5. Tronconi, M. *MEMS and Sensors Are the Key Enablers of Internet of Things*; STMicroelectronics: Geneva, Switzerland, 2013.
6. Hernandez, J.; Li, Y.; Rehg, J.M.; Picard, R.W. BioGlass: Physiological Parameter Estimation Using a Head-mounted Wearable Device. In Proceedings of the Wireless Mobile Communication and Healthcare, Athens, Greece, 3–5 November 2014; pp. 55–58.
7. Pierleoni, P.; Belli, A.; Palma, L.; Pellegrini, M. A High reliability wearable device for elderly fall detection. *IEEE Sens. J.* **2015**, *15*, 4544–4553. [CrossRef]
8. Bernstein, J.; Cho, S.; King, A.T.; Kourepenis, A.; Maciel, P.; Weinberg, M. A micromachined comb-drive tuning fork rate gyroscope. In Proceedings of the IEEE Micro Electro Mechanical Systems, Fort Lauderdale, FL, USA, 10 February 1993; pp. 143–148.
9. Yang, C.; Li, H. Digital control system for the MEMS tuning fork gyroscope based on synchronous integral demodulator. *IEEE Sens. J.* **2015**, *15*, 5755–5764. [CrossRef]

10. Varadan, V.K.; Suh, W.D.; Xavier, P.B.; Jose, K.A.; Varadan, V.V. Design and development of a MEMS-IDT gyroscope. *Smart Mater. Struct.* **2000**, *9*, 898–905.
11. Zhang, L.; Masek, V.; Sanatdoost, N.N. Structural optimization of Z-axis tuning-fork MEMS gyroscopes for enhancing reliability and resolution. *Microsyst. Technol.* **2014**, *21*, 1187–1201. [CrossRef]
12. Yazdi, N.; Ayazi, F.; Najafi, K. Micromachined inertial sensors. *Proc. IEEE* **1998**, *86*, 1640–1659. [CrossRef]
13. Faculty, T.A.; Fulfillment, I.P. Degree-Per-Hour Mode-Matched Micromachined Silicon Vibratory Gyroscopes. Ph.D. Thesis, Georgia Tech School of Mechanical Engineering, Atlanta, GA, USA, 2008.
14. Merlo, S.; Norgia, M.; Donati, S. Fiber gyroscope principles. In *Handbook of Fibre Optic Sensing Technology*; John Wiley & Sons Ltd.: New York, NY, USA, 2002; pp. 1–23.
15. Yu, Y.Y.; Chen, B.Y.; Tao, J.; Chen, X.J.; Zhang, H.; Pang, W.; Zhang, D.H.; Luo, H. A novel high sensitivity MEMS acoustic gyroscope by measuring phase shift. In Proceedings of the IEEE Sensors, Busan, Korea, 1–4 November 2015.
16. Sanders, G.A.; Sanders, S.J.; Strandjord, L.K.; Qiu, T.; Wu, J.; Smiciklas, M.; Salit, M. Fiber optic gyroscope development at Honeywell. In Proceedings of the SPIE Commercial+ Scientific Sensing and Imaging, International Society for Optics and Photonics, Baltimore, MD, USA, 14 May 2016.
17. Song, N.; Cai, W.; Song, J.; Jin, J.; Wu, C. Structure optimization of small-diameter polarization-maintaining photonic crystal fiber for mini coil of space borne miniature fiber-optic gyroscope. *Appl. Opt.* **2015**, *54*, 9831–9838. [CrossRef] [PubMed]
18. Ciminelli, C.; Dell'Olio, F.; Armenise, M.N. High-Q spiral resonator for optical gyroscope applications: numerical and experimental investigation. *IEEE Photonics J.* **2012**, *4*, 1844–1854. [CrossRef]
19. Lu, Y.; Shelton, S.; Horsley, D.A. High frequency and high fill factor piezoelectric micromachined ultrasonic transducers based on cavity SOI wafers. In Proceedings of the Solid-State Sensors, Actuators, and Microsystems Workshop, Hilton Head Island, SC, USA, 8–12 June 2014; pp. 131–134.
20. Lu, Y.; Heidari, A.; Horsley, D.A. A high fill-factor annular array of high frequency piezoelectric micromachined ultrasonic transducers. *J. Microelectromech. Syst.* **2015**, *24*, 904–913. [CrossRef]
21. Przybyla, R.; Izyumin, I.; Kline, M.; Boser, B.; Shelton, S. An ultrasonic rangefinder based on an AlN piezoelectric micromachined ultrasound transducer. In Proceedings of the IEEE Sensors, Waikoloa, HI, USA, 1–4 November 2010; pp. 2417–2421.
22. Wygant, I.O.; Kupnik, M.J.; Windsor, J.C.; Wright, W.M.; Wochner, M.S.; Hamilton, M.F. 50 kHz capacitive micromachined ultrasonic transducers for generation of highly directional sound with parametric arrays. *IEEE Trans. Ultrason. Ferroelectr. Freq. Control* **2009**, *56*, 193–203. [CrossRef] [PubMed]
23. Shelton, S.; Chan, M.L.; Park, H.; Horsley, D.A. CMOS-compatible AlN piezoelectric micromachined ultrasonic transducers. In Proceedings of the IEEE International Ultrasonics Symposium, Rome, Italy, 20–23 September 2009; pp. 402–405.
24. Lu, Y.; Heidari, A.; Shelton, S.; Guedes, A.; Horsley, D.A. High frequency piezoelectric micromachined ultrasonic transducer array for intravascular ultrasound imaging. In Proceedings of the IEEE 27th MEMS, San Francisco, CA, USA, 26–30 January 2014; pp. 745–748.
25. Muralt, P.; Ledermann, N.; Paborowski, J.; Barzegar, A.; Gentil, S.; Belgacem, B.; Petitgrand, S.; Bosseboeuf, A.; Setter, N. Piezoelectric micromachined ultrasonic transducers based on PZT thin films. *IEEE Trans. Ultrason. Ferroelectr. Freq. Control* **2005**, *52*, 2276–2288. [CrossRef] [PubMed]
26. Shelton, S.; Rozen, O.; Guedes, A.; Przybyla, R.; Boser, B.; Horsley, D.A. Improved acoustic coupling of air-coupled micromachined ultrasonic transducers. In Proceedings of the 2014 IEEE 27th International Conference on Micro Electro Mechanical Systems, San Francisco, CA, USA, 26–30 January 2014; pp. 753–756.
27. Sammoura, F.; Smyth, K.; Kim, S.-G. Optimizing the electrode size of circular bimorph plates with different boundary conditions for maximum deflection of piezoelectric micromachined ultrasonic transducers. *Ultrasonics* **2013**, *53*, 328–334. [CrossRef] [PubMed]
28. Lu, Y.; Horsley, D.A. Modeling, fabrication, and characterization of piezoelectric micromachined ultrasonic transducer arrays based on cavity SOI wafers. *J. Microelectromech. Syst.* **2015**, *24*, 1142–1149. [CrossRef]
29. Horsley, D.A.; Rozen, O.; Lu, Y.; Shelton, S.; Guedes, A.; Przybyla, R.; Tang, H.Y.; Boser, B.E. Piezoelectric micromachined ultrasonic transducers for human-machine interfaces and biometric sensing. In Proceedings of the IEEE Sensors, Melbourne, Australia, 2–5 November 2015.
30. Muralt, P.; Baborowski, J. Micromachined ultrasonic transducers and acoustic sensors based on piezoelectric thin films. *J. Electroceram.* **2004**, *12*, 101–108. [CrossRef]

31. Li, X.; Xu, L.; Sun, T. Piezoelectric micromachined ultrasonic transducer array for micro audio directional loudspeaker. In Proceedings of the IEEE International Conference on Mechatronics and Automation, Takamatsu, Japan, 4–7 August 2013; pp. 450–455.

32. Qiu, Y.Q.; Gigliotti, J.V.; Wallace, M.; Griggio, F.; Demore, C.E.M.; Cochran, S.; McKinstry, T.S. Piezoelectric micromachined ultrasound transducer (PMUT) arrays for integrated sensing, actuation and imaging. *Sensors* **2015**, *15*, 8020–8041. [CrossRef] [PubMed]

33. Przybyla, R.; Flynn, A.; Jain, V.; Shelton, S.; Guedes, A.A.; Izyumin, I.; Horsley, D.; Boser, B.E. A micromechanical ultrasonic distance sensor with > 1 meter range. In Proceedings of the 16th International Solid-State Sensors, Actuators and Microsystems Conference (TRANSDUCERS), Beijing, China, 5–9 June 2011; pp. 2070–2073.

34. Shelton, S.; Guedes, A.; Przybyla, R.; Tsai, J.M. Aluminum nitride piezoelectric micromachined ultrasound transducer arrays. In Proceedings of the Solid-State Sensors, Actuators, and Microsystems Workshop, Hilton Head, SC, USA, 11–14 June 2012; pp. 291–294.

35. Kinsler, L.E.; Fery, A.R.; Coppens, A.B.; Sanders, J.V. *Fundamentals of Acoustics*; Wiley: Hoboken, NJ, USA, 1999.

36. Blackstock, D. *Fundamentals of Physical Acoustics*; John Wiley & Sons: Hoboken, NJ, USA, 2000.

37. Tang, H.; Hu, Y.; Fung, S.; Tsai, J.M. Pulse-echo ultrasonic fingerprint sensor on a chip. In Proceedings of the 18th Solid-State Sensors, Actuators and Microsystems (TRANSDUCERS), Anchorage, AK, USA, 21–25 June 2015; pp. 674–677.

38. *ANSI/IEEE Std 176-1987, IEEE Standard on Piezoelectricity*; The Institute of Electrical and Electronics Engineers, Inc.: New York, NY, USA, 1988.

39. Sammoura, F.; Shelton, S.; Akhbari, S.; Horsley, D.; Lin, L.L. A Two-Port Piezoelectric Micromachined Ultrasonic Transducer. In Proceedings of the 2014 Joint IEEE International Symposium on the Applications of Ferroelectrics, International Workshop on Acoustic Transduction Materials and Devices & Workshop on Piezoresponse Force Microscopy (ISAF/IWATMD/PFM), State College, PA, USA, 1–4 May 2014.

40. Geng, Y.L.; Xu, L.M.; Wang, Y. Optimization of a circular piezoelectric micro-mechanical ultrasonic transducer on electromechanical coupling coefficient. In Proceedings of the Piezoelectricity, Acoustic Waves, and Device Applications (SPAWDA) and 2009 China Symposium on Frequency Control Technology, Wuhan, China, 17–20 December 2009.

41. Jung, J.; Kim, S.; Lee, W.; Choi, H. Fabrication of a two-dimensional piezoelectric micromachined ultrasonic transducer array using a top-crossover-to-bottom structure and metal bridge connections. *J. Micromech. Microeng.* **2013**, *23*, 125037. [CrossRef]

micromachines

MDPI

Article

Spiral-Shaped Piezoelectric MEMS Cantilever Array for Fully Implantable Hearing Systems

Péter Udvardi [1], János Radó [1,2], András Straszner [1], János Ferencz [1], Zoltán Hajnal [1], Saeedeh Soleimani [1], Michael Schneider [3], Ulrich Schmid [3], Péter Révész [4] and János Volk [1,*]

[1] Institute for Technical Physics and Materials Science, MTA EK, 1121 Konkoly Thege M. út 29-33, H-1121 Budapest, Hungary; udvardi.peter98@gmail.com (P.U.); rado@mfa.kfki.hu (J.R.); straszner@mfa.kfki.hu (A.S.); ferencz@mfa.kfki.hu (J.F.); hajnal@mfa.kfki.hu (Z.H.); soleimani.saeedeh@energia.mta.hu (S.S.)
[2] Doctoral School on Material Sciences and Technologies, Óbuda University, Bécsi út 96/b, H-1034 Budapest, Hungary
[3] Institute of Sensor and Actuator Systems, TU Wien, 1040 Vienna, Austria; michael.schneider@tuwien.ac.at (M.S.); ulrich.e366.schmid@tuwien.ac.at (U.S.)
[4] Department of Otorhinolaryngology-Head and Neck Surgery, Clinical Center, University of Pécs, H-7601 Pécs, Hungary; revesz.peter@pte.hu
* Correspondence: volk@mfa.kfki.hu; Tel.: +36-1-392-2222 (ext. 3474)

Received: 6 September 2017; Accepted: 12 October 2017; Published: 18 October 2017

Abstract: Fully implantable, self-powered hearing aids with no external unit could significantly increase the life quality of patients suffering severe hearing loss. This highly demanding concept, however, requires a strongly miniaturized device which is fully implantable in the middle/inner ear and includes the following components: frequency selective microphone or accelerometer, energy harvesting device, speech processor, and cochlear multielectrode. Here we demonstrate a low volume, piezoelectric micro-electromechanical system (MEMS) cantilever array which is sensitive, even in the lower part of the voice frequency range (300–700 Hz). The test array consisting of 16 cantilevers has been fabricated by standard bulk micromachining using a Si-on-Insulator (SOI) wafer and aluminum nitride (AlN) as a complementary metal-oxide-semiconductor (CMOS) and biocompatible piezoelectric material. The low frequency and low device footprint are ensured by Archimedean spiral geometry and Si seismic mass. Experimentally detected resonance frequencies were validated by an analytical model. The generated open circuit voltage (3–10 mV) is sufficient for the direct analog conversion of the signals for cochlear multielectrode implants.

Keywords: artificial basilar membrane; cochlear implant; frequency selectivity; Archimedean spiral; aluminum nitride (AlN); piezoelectric cantilever; micro-electromechanical system (MEMS); finite element analysis; energy harvesting

1. Introduction

Cochlear implant (CI) is a surgically implanted electronic device that provides a sense of sound to a person who suffers from profound hearing loss or deafness. The present generation of hearing systems bypass the normal hearing process. Outside the skin, it consists of a microphone, a speech processor, and a transmitter, which transmits signals through an internal receiver to an array of electrodes placed in the cochlea. Though, in the last three decades the technology has undergone an impressive improvement, there are still some challenges to be addressed for higher wearing comfort [1]. Because of the external units, the system is visible, making the patients stigmatized. It also limits the activities that can be undertaken while wearing the device. Broken wires, cables, and speech processors can cause derangement, too. This could be minimized by having a fully implantable cochlear implant (FICI, sometimes also referred to as totally implantable cochlear implant, TICI) which functions round-the-clock while sleeping, showering, swimming, and during many types of vigorous physical activities [2]. FICI is supposed to be composed of an internal microphone or a piezoelectric acoustic sensor, an electronic device which transform the signal for the hearing nerves, a battery and/or an energy harvesting unit, as well as a multielectrode array inserted in the cochlea.

Several solutions have been proposed to mimic the frequency selectivity (tonotopy) of the cochlea. The topologically more faithful physical model is an elastic membrane having varying widths along its length [3–7]. The other approach is to apply an array of micro-electromechanical system (MEMS) cantilevers having varying length, and thus, varying natural resonance frequency. Because of the smaller size and more reliable fabrication procedure, the latter approach seems to have higher potential for FICI. In 1998 Harada et al. demonstrated a fishbone structured acoustic sensor using piezoresistive read-out elements [8]. Later, Xu et al. reported polymeric micro-cantilever array to mimic the mammalian cochlea [9]. Recently, Jang et al. used an aluminum nitride (AlN) coated array as an artificial basilar membrane [10,11]. Though the rectangular unimorph cantilevers showed excellent frequency selectivity, they covered only the upper half of the human hearing range (2.9–12.6 kHz) [11], since the natural resonance frequency at fixed cantilever thickness scales up with decreasing length. To obtain sensitive cantilevers in the range of 300–700 Hz is, however, more challenging.

Low frequency, spiral, and spiral-like cantilevers were proposed and theoretically evaluated by Choi et al. in 2006 [12], and recently, also experimentally demonstrated. Zhang et al. [13], and Lu et al. [14] reported an S-shaped Lead Zirconate Titanate (PZT) coated flexure suspended MEMS device on a chip size of 6 mm × 6 mm, for vibration sensing and energy harvesting. Though the covered frequency range is attractive for the proposed FICI concept, a smaller footprint, higher Q-factor (>100), and bio- and metal-oxide-semiconductor (CMOS) compatible piezoelectric material, like AlN [15,16], are needed for the device to be implantable in the human middle ear.

Voice detection in CIs is done directly by perceiving the modulation of air or fluid pressure using a microphone, even if the functions of the tympanic membrane and of the middle ear are intact. An alternative approach, also an aim in this paper, is to measure the vibration of the ossicles with a miniaturized MEMS based implant. As a guideline for the design, we referred to the work of Gan et al. [17] on implantable middle ear hearing devices (IMEHDs). In their work, a small magnet was mounted between the malleus and stapes, and driven electromagnetically by a coil placed under the ear canal bony wall. The diameter and length of the cylinder-shaped magnet is 1.5 and 2 mm, respectively, and it weighs 26 mg, which is comparable to the size and mass of a 3D packaged Si multicantilever system. Beker et al. demonstrated a rectangular cantilever with Si seismic mass to achieve resonance in this lower voice frequency range (474 Hz) [18]. However, the size (6 mm × 6 mm) and weight of the chip, especially in packaged multichannel form, is too large to be fixed onto one of the middle ear bones. Moreover, the applied technique to bond bulk piezoceramic PZT dices to the substrate, and the grinding, make the wafer scale processing highly demanding.

Here, we demonstrate an array of spiral cantilevers with Si seismic mass at their ends, which is optimized to achieve a small footprint, compactness, bio- and CMOS compatibility, low resonance frequency (300–700 Hz), high Q-factor in air (117–254), and high robustness with limited internal stress in the Si cantilever. These 2 mm × 2 mm cantilever chips are small enough to pack them in a compact multichannel device which fits into the middle auditory system, and can provide a new solution for next generation FICI systems.

2. Materials and Methods

2.1. Cantilever Design

According to the Euler–Bernoulli theory, the first natural frequency ($n = 1$) of a one-side-clamped rectangular Si cantilever is inversely proportional to the square of the length (L) of the beam [19]:

$$f_{\text{beam}} = \frac{\beta_1^2}{2\pi} \sqrt{\frac{E_{\text{Si}}}{12\rho_{\text{Si}}} \frac{t}{L^2}}, \tag{1}$$

where $\beta_1 = 1.875$, t is thickness of the cantilever, E_{Si} is the Young modulus for <110> crystallographic orientation (1.69×10^{11} Pa) [20] and ρ_{Si} is the mass density (2330 kg/m^3) of Si. It means that using a Si-on-Insulator (SOI) wafer, with a typical Si device thickness of 12.5 μm and a cantilever length of 7.6 mm, is needed to reach the lower end of the voice frequency range (300 Hz). This size is too large to allow implantation of the device in the inner ear. However, by applying a tip mass (M) at the free end of the beam with a distributed mass (m), the first natural frequency can be reduced as follows [21]:

$$f_{\text{tm}} = \frac{1}{2\pi} \sqrt{\frac{3E_{\text{Si}}Wt^3}{12(M+0.24m)L^3}}. \tag{2}$$

If we assume that the tip mass is significantly higher than that of the cantilever ($M \gg m$) the reduced frequency due to the tip mass can be approximated by

$$f_{\text{tm}} = \sqrt{\frac{m}{M}} f_{\text{beam}}. \tag{3}$$

Moreover, the seismic mass also helps to minimize the air damping effect, and increases the stored energy [12]. However, too large a mass results in a high internal stress, and a fracture of the cantilever upon resonance. Therefore, a trade-off is needed between cantilever length and the proof mass.

The one side clamped spiral cantilevers were designed to fit into a 2 mm × 2 mm square window. The thickness of the single crystal Si beam is 12.5 μm, which corresponds to the device layer of the selected SOI wafer. The arc of the beam was defined by the parametric equation of an Archimedean spiral, where the curvature radius is continuously decreasing, with the azimuth angle φ from an initial radius at the clamping point (R_0), until reaching the radius of the proof mass (r_0),

$$\begin{pmatrix} x \\ y \end{pmatrix} = (R_0 - c\varphi) \begin{pmatrix} \cos(\varphi) \\ \sin(\varphi) \end{pmatrix}. \tag{4}$$

Finite element analysis, using COMSOL Multiphysics (5.2a, Burlington, MA, USA), was used to select 16 different geometries which fulfilled the following two requirements: (i) the (first) significant natural resonance has to fall in the frequency range of 300–1000 Hz; and (ii) the maximal von Mises stress along the cantilever at a driving acceleration of 1 g is not allowed to exceed 5% of the fracture strength of anisotropically etched Si diaphragms, i.e., 300 MPa [22]. One typical stress distribution in Figure 1a shows that the stress is gradually decreasing from the clamping side toward the tip mass without reaching the chosen limit of 15 MPa. Several spiral geometries were tested numerically by changing the parameters as follows: width of the cantilever in the range of 140–180 μm, the number of the turns between 3 and 4 φ = 6–8π), c parameter in a range of 27.0–36.5 μm/rad, and the radius of the tip mass between 160 to 300 μm. The selected 16 spirals are shown in Figure 1b.

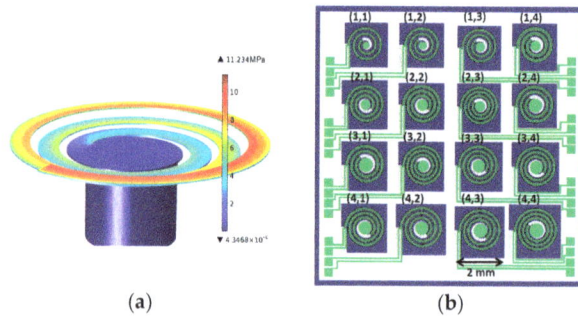

(a)   (b)

Figure 1. (a) Tensile stress distribution along the beam calculated by finite element analysis to exclude "fragile" geometries upon selection; **(b)** Layout of the selected 4 × 4 spirals. Each of the spirals fit into a square area of 2 mm × 2 mm. Five geometrical parameters of the Archimedean spirals were varied to tune the natural frequency: number of the half turns, width (W), starting radius at the clamping point (R_0), radius of the proof mass cylinder (r_0), and c parameter, describing the rate of the radius change from the edge towards the center.

Since the polarization axis of AlN thin films (c-axis) is nearly perpendicular to the substrate, a d_{31} type contacting scheme was used (top-bottom) to generate electrical signals. The piezo layer does not fully cover the whole length of the arc, as the last segment towards the seismic mass is almost stress-free.

2.2. Analytical Model

The model was based on the one described by Karami et al. [23], with a few modifications to include the effects of the tip mass attached to the spiral cantilever. Therefore, the kinetic energy T and the potential energy V for the system are

$$T = \int \frac{1}{2}\rho A \left(\frac{\partial w}{\partial t}\right)^2 + \frac{1}{2}I_z\left(\frac{\partial \beta}{\partial t}\right)^2 dx + \frac{1}{2}M\left(\frac{\partial w}{\partial t}\right)^2\Big|_{x=L} + \frac{1}{2}I_{t_x}\left(\frac{\partial^2 w}{\partial x \partial t}\right)^2\Big|_{x=L} + \frac{1}{2}I_{t_z}\left(\frac{\partial \beta}{\partial t}\right)^2\Big|_{x=L} \quad (5)$$

$$V = \int \frac{1}{2}\left[EI\left(\frac{\beta}{R} - \frac{\partial^2 w}{\partial x^2}\right)^2 + GJ\left(\frac{\partial \beta}{\partial x} + \frac{1}{R}\frac{\partial w}{\partial x}\right)^2\right]dx \quad (6)$$

where $w(x,t)$ is the deflection, $\beta(x,t)$ is the twist angle of the cantilever, $R(x)$ is the radius of the spiral, A is the area of the cantilever, E is the composite Young modulus, I is the second moment of inertia, J is the torsion constant, G is the shear modulus (modulus of rigidity), I_{t_x} and I_{t_z} are the mass moment of inertia of the seismic mass with respect to x and z axis. To account for the AlN (E_{AlN} = 344 GPa) and Au (E_{AlN} = 79 GPa) [24] layers deposited on top of the Si, the Young's modulus in the calculations was taken for the composite material as described in Timoshenko and Young's textbook [25,26]. For the thicker (12.5 μm) Si membrane, the crystal orientation dependence of Si can also play a role. Since the in-plane orientation of the spiral alternate between <100> and <110> directions along the spiral arc with an angular periodicity of $\pi/2$, the average of the corresponding Young moduli (130 GPa and 169 GPa, respectively [26]) was taken into account in the model. Thus, the obtained composite Young modulus for Equation (6) was E = 212 GPa.

Equating the variation of the action ($\int T - V dt$) to be zero yields six equations, from which only four constrain the system on the spatial boundaries, and the solutions are only functions of time.

Hence, transforming all six equations into frequency domains yields two coupled equations for the spatial coordinates, which are as follows,

$$\omega^2 \rho A w + \frac{d^2}{dx^2} EI \left[\frac{\beta}{R} + \frac{d^2 w}{dx^2} \right] + \frac{d}{dx} \frac{GJ}{R} \left[\frac{d\beta}{dx} + \frac{1}{R} \frac{dw}{dx} \right] = \omega^2 \rho A w_b, \tag{7}$$

$$-\omega^2 I_z \beta + \frac{EI}{R} \left[\frac{\beta}{R} - \frac{d^2 w}{dx^2} \right] - \frac{d}{dx} GJ \left[\frac{d\beta}{dx} + \frac{1}{R} \frac{\partial w}{dx} \right] = 0, \tag{8}$$

as well as four equations, which provide natural boundary conditions for the equations above:

$$\left[-\omega^2 M w + \frac{d}{dx} EI \left[\frac{\beta}{R} - \frac{d^2 w}{dx^2} \right] + \frac{GJ}{R} \left[\frac{d\beta}{dx} + \frac{1}{R} \frac{dw}{dx} \right] \right]_{x=L} = 0 \tag{9}$$

$$\left[EI \left[\frac{\beta}{R} - \frac{d^2 w}{dx^2} \right] + \omega^2 I_{t_x} \frac{dw}{dx} \right]_{x=L} = 0 \tag{10}$$

$$\left[-\omega^2 I_{t_z} \beta + GJ \left[\frac{d\beta}{dx} + \frac{1}{R} \frac{dw}{dx} \right] \right]_{x=L} = 0 \tag{11}$$

$$w(0) = \frac{dw}{dx}|_{x=0} = \beta(0) = 0 \tag{12}$$

The equations were solved using the Chebyshev spectral collocation method, implemented by the library of the open-source software, Chebfun. The library was chosen as it provides a quick solution, even for a high-resolution frequency sweep.

2.3. Fabrication

The unimorph, d_{31} type piezocantilever arrays were fabricated on a 4″ SOI wafer with 12.5 μm Si device layer, 1 μm buried oxide (BOX), and a handle layer of 550 μm. The fabrication process is shown in Figure 2. At first, thermal oxide layer with a thickness of 300 nm was grown onto the wafer (Figure 2B). The bottom electrode of Ti (30 nm)/Au (120 nm) was prepared by e-beam evaporation (AJA) and a subsequent lift-off step (Figure 2C). The piezoelectric AlN layer, having a thickness of 830 nm, was deposited by reactive radio frequency (RF) sputtering from an 8″ Al target in a Leibold Heraeus Z550 system (CAE Inc., Montreal, QC, Canada). No additional substrate heating was applied beyond the natural effect of the RF generated nitrogen plasma (500 W). The maximum substrate temperature during the deposition is around 300 °C. The AlN layer was then patterned by photolithography and wet chemical etching using phosphoric acid (Figure 2D). It was followed by the deposition of a Ti (30 nm)/Au (120 nm) top electrode at the same conditions used for its lower counter electrode (Figure 2E). The micromachining of the spiral cantilevers was started from the front side using a deep reactive ion etching (DRIE) step in an Oxford Plasmalab System 100 ICP180 (Oxford Instruments Plasma Technologies, Yatton, Bristol, UK). The etching was performed through the whole device layer of 12.5 μm, and was stopped by the BOX layer (Figure 2F). Etching from the back side was carried out by Bosch process, in such a way to obtain full wafer thick (550 μm) seismic masses at the free end of the cantilevers. In addition, it was split into two steps using, at first, photoresist, and then Al mask, with slightly different patterns to obtain perforated Si wafers, in which the 4 × 4 block arrays are connected by thinned (~150 μm) Si bridges (Figure 2G,H). The aim of this method was to avoid chip dicing, which would have damaged the sensitive cantilevers. The wafer process was followed by etching of the Al hard mask in phosphoric acid (Figure 2I), and finished by the wet etching of the buried oxide in hydrofluoric acid, to release the spiral cantilevers (Figure 2J). The manually cleaved dices were firmly mounted, and wire bonded onto a printed circuit board (PCB) recessed under the cantilever, which ensured the electrical readout and the free vibration of the piezo cantilevers.

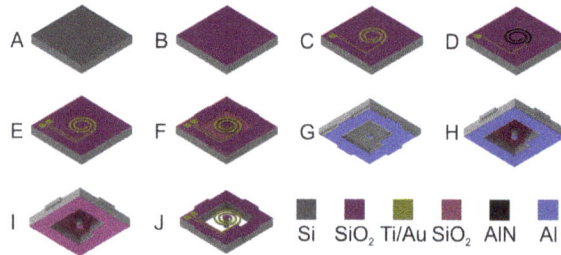

Figure 2. Schematics of the fabrication procedure: (**A**) 4″ Si-on-Insulator (SOI) wafer; (**B**) thermal oxidation (300 nm); (**C**) deposition and lift-off of the bottom Ti/Au contact and pads; (**D**) radio frequency (RF) sputter deposition and patterned etching of aluminum nitride (AlN) (830 nm); (**E**) deposition and lift-off of top Ti/Au contact; (**F**) deep reactive ion etching (DRIE) of the spiral beam from the front side; (**G**) first DRIE Bosch process from the back-side and the strip of the photoresist mask; (**H**) second DRIE Bosch process from the back-side through the Al hard mask with a slightly modified pattern to obtain perforated Si frame; (**I**) etching of the Al masking layer; (**J**) hydrofluoric acid (HF) etching of the buried oxide and removal of the top protective photoresist layer.

2.4. Characterization

The fabricated piezoelectric cantilevers were analyzed in a LEO XB-1540 crossbeam scanning electron microscope (SEM) (Zeiss, Oberkochen, Germany) and in a New View 7100 optical surface profiler (ZYGO, Middlefield, CT, USA). During the electromechanical tests, the PCB (Figure 3a) was placed on a purpose designed 3D printed chip holder mounted with a miniaturized calibrated accelerometer (4397-A, Brüel & Kjaer, Nærum, Denmark) (Figure 3b). Vibrations were carried out by a mini shaker (LDS V201) which was driven by a power amplifier (2735, Brüel & Kjaer) controlled by signal generator (AFG 3252C, Tectronix, Beaverton, OR, USA) (Figure 3c). LabView software was written to perform frequency sweeps and collect open circuit voltage (V_{OC}) signals through a data acquisition card (USB DAQ 6211, National Instruments, Austin, TX, USA). The amplitude of the generated sinusoidal signal was controlled by a closed feedback loop mechanism using the signal or the accelerometer. In this way, the acceleration is fixed to a constant level during the frequency sweep. The output voltage signals were collected for each cantilever at continuous sweep of the sinusoidal excitation in the frequency range of 20 Hz–1.2 kHz. It is worth mentioning that in contrast to several studies, we did not measure the direct effect of sound pressure wave on the sensor. Instead, we applied an external acceleration on the frame of the Si chip.

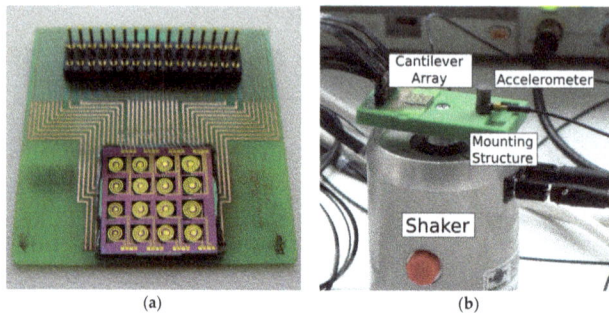

(a)

(b)

Figure 3. *Cont.*

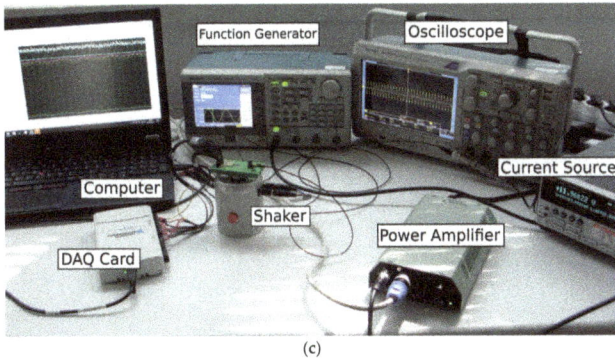

(c)

Figure 3. Experimental setup for the characterization of the piezoelectric cantilever arrays: (**a**) Wire bonded cantilever array on the printed circuit board (PCB); (**b**) 3D printed sample holder with a calibrated accelerometer mounted on a shaker; (**c**) the shaker was controlled by a function generator through a power amplifier. The output voltage signals of the cantilevers and the calibrated accelerometer were collected by a programmed data acquisition card (DAQ). The signals were in situ visualized by an oscilloscope during the automatized frequency sweeps. Current source was used to feed the accelerometer.

3. Results and Discussion

3.1. Structural Characterization

Most of the chips on the 4″ wafer with the sensitive spirals survived the over 30-step fabrication process; an example is shown in Figure 3a. It was especially critical when releasing the cantilevers from the buried SiO_2 membrane (Figure 2J) and during the manual dicing of the cantilever arrays. The cantilevers also tolerated the vibration test up to an instrumental acceleration limit of 5 g. Scanning electron micrograph of two typical spiral cantilevers situated in the (1,3) and (3,4) array positions are shown in Figure 4a,b, respectively. The darker region on the cantilever beam corresponds to the metal–piezo–metal stack covered region. A reflecting Au disk in the center was designed for additional laser beam deflection tests. As shown in both SEM images, the diameter of the seismic mass is decreasing from the back side towards the membrane, i.e., the sidewall of the DRIE etching was not perpendicular. By a closer SEM observation, the cone angle was found to be 7°.

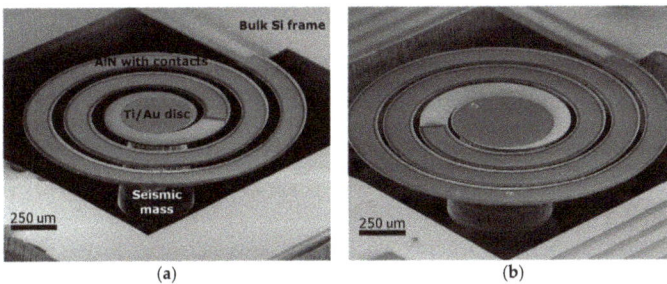

(a)   (b)

Figure 4. Tilt-view scanning electron microscope (SEM) images of two typical suspended spiral-shaped cantilevers (1,3) (**a**) and (3,4) (**b**) with contacted AlN layer (darker region) on their top surfaces, and a wafer thick 3D-micromachined Si seismic mass beneath. The darker circle in the center corresponds to a Ti/Au disc applied for laser reflection tests. Truncated shape of the tip mass is due to the increasing underetching ratio during the DRIE.

AlN thin films often have a significant compressive or tensile residual stress which depends on several factors, such as the deposition technique (RF sputtering, pulsed DC sputtering, chemical vapor deposition (CVD), etc.), growth parameters (flow rate, deposition temperature, plasma power, etc.) [27], the material quality of the underlaying template [28], or the layer thickness [29]. In our previous studies, we found that AlN layers having a similar thickness deposited directly on Si wafer in our RF sputtering system resulted in a significant tensile stress (~575 MPa) in the layer. In contrast, the freely suspended spiral cantilevers seem to be flat in the SEM images (Figure 4); i.e., neither significant downward deflection due to gravity, nor upward deformation due to tensile stress, is visible. Nevertheless, in order to quantify the strain, we performed tests of optical surface profile on two cantilevers. The first one was an intact spiral of an array position of (1,1) with tip mass. From the second chip (4,3), the tip mass was intentionally removed to study the effect of the residual stress directly. Figure 5a,b show the corresponding 3D images recorded on one side of the spirals by a 10× objective lens. Height line profiles taken along the dashed lines for the cantilever, with and without tip mass, are shown in Figure 5c,d, respectively. As shown in Figure 5a, though the tip mass is tilted due to its torsional moment, the height difference between the clamping point and in the same y position in the second turn is 1.0 μm, though the arc length between the two points is about 6 mm. Without tip mass, the deflection is even smaller, and the height changes around 0.3 μm along an arc length of about 9.3 mm. This indicates that the stress in AlN on the spiral is negligible, and the deflection is mainly affected by the static load of the central seismic mass. This low residual stress, compared to our previously found results, can be attributed either to the beneficial effect of the Au bottom contact upon sputter deposition, or to the device geometry.

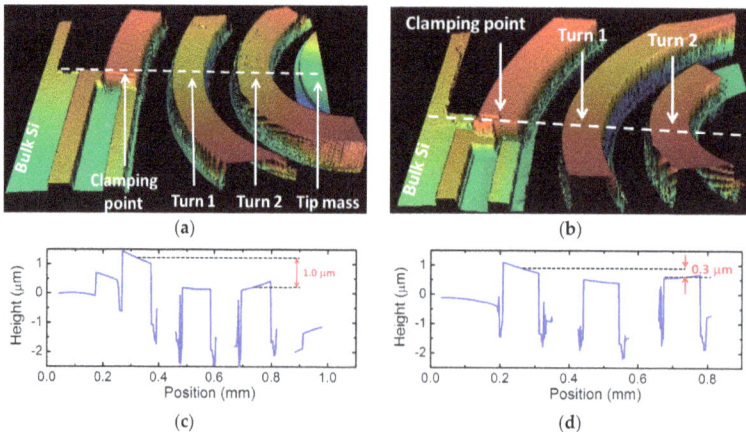

Figure 5. 3D optical surface profiler images taken on a portion of a suspended spiral with (**a**) and without (**b**) seismic mass at its end. Characteristic height profiles along the dashed lines for cantilever with (**c**) and without seismic mass (**d**).

3.2. Vibration Tests and Validation of the Analytical Model

Vibration tests, performed at a fixed acceleration of 1 g (9.81 m/s^2), confirmed the frequency selectivity of the cantilever array (Figure 6). Most of the spirals (12 out 16) showed sharp resonance peaks scattering in the range of 281–672 Hz. For clarity, the spectra were ordered by their resonance frequencies from low to high (Channel 1–12). The calculated Q-factors ($f_0/\Delta f_{\sqrt{2}}$) in air vary in the range of 117–254. As shown in the inset image of Figure 5, the recorded time-dependent open circuit signal is purely sinusoidal, without any vertical offset, which indicates stress-free cantilevers. The inactivity of the remaining four devices can be attributed to the low electrical resistance of the

metal/AlN/metal stacks, which may be the results of random defects in the AlN layer. The generated
open circuit voltages fell in the range of 3.0–9.6 mV.

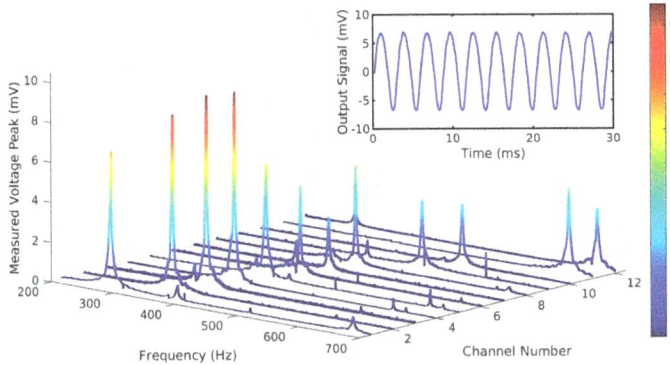

Figure 6. Piezoelectric output voltage during continuous sinusoidal excitation at a feedback controlled
acceleration of 1 g. Depending on the geometry, the base frequency of the cantilevers falls in the range
of 281–673 Hz. Inset shows the sinusoidal output waveform of channel 1 at resonance.

Using the analytical method described in Section 2.2, resonance frequencies were calculated
for each cantilever by taking into account the tapered geometry of the seismic mass, and the layer
stack on the top of the cantilever. The obtained frequency values have good agreement with the
experimental ones; the calculated root-mean-square deviation is 21.5 Hz. Table 1 summarizes the
geometrical parameters, the experimentally and numerically obtained resonance frequencies, as well
as the generated V_{OC} amplitude for each cantilever.

Table 1. Geometrical parameters, calculated and measured resonance frequencies, peak output voltages,
quality factors for each piezoelectric cantilever, as well as the resonance frequency of uncoated spiral
beam and the corresponding theoretical values.

Ch. No.	L	w	R_0	r	m_{beam}	m_{tm}	Res. Freq. Exp.	Res. Freq. Calc.	Peak Voltage	Q Factor	Res. Fr./Etched Exp.	Res. Fr./Etched Calc.
	(π/mm)	(μm)	(μm)	(μm)	(ng)	(ng)	(Hz)	(Hz)	(mV)		(Hz)	(Hz)
1	8.0/11.8	145	870	315	50	320	281	283	6.89	145	241	240
2	7.0/9.4	150	850	300	41	287	345	337	8.80	133	289	295
3	7.0/9.2	150	800	300	40	287	365	366	9.60	148	311	312
4	7.0/9.2	150	800	290	40	266	374	348	9.53	254	321	319
5	7.0/9.1	145	800	280	38	246	389	377	5.71	117	335	332
6	6.5/7.4	160	760	310	36	309	408	379	4.47	218	343	348
7	8.0/10.3	145	800	250	40	190	417	424	2.64	176	352	357
8	6.0/7.4	160	760	300	33	287	425	473	5.02	216	-[2]	362
9	7.0/7.7	160	760	250	36	190	497	471	3.40	180	422	424
10	7.0/7.7	150	760	230	34	157	526	543	3.05	173	449	449
11	6.5/6.3	180	700	230	34	157	663	679	4.40	245	578	576
12	6.0/6.5	140	700	210	27	127	672	687	3.14	210	590	574
13	7.0/9.9	145	850	310	42	309	1	363	-	-	268	-
14	8.0/9.6	140	760	230	40	157	1	482	-	-	414	-
15	6.0/6.6	180	700	180	27	88	1	874	-	-	707	-
16	6.0/6.4	140	700	160	26	66	1	1058	-	-	2	-

[1] No resonance frequency was found in the 200–1200 Hz range. [2] Cantilever was broken during etching.

In order to validate the analytical model, and to see if any residual stress in the coating causes
systematic shift in the resonance frequency, we performed vibration tests on coating free bare
cantilevers. For that we removed the SiO_2/Ti/Au/AlN/Ti/Au stacks from one chip in subsequent

etching steps of potassium iodide solution, hot phosphoric acid (85 °C), potassium iodide solution, and hydrofluoric acid. The resonance frequencies of the bare non-piezoelectric cantilevers were determined by eye under optical microscope at manual frequency sweeps. As expected, the obtained natural resonance frequencies are lower than the corresponding values for coated cantilevers. However, it is a question if this 14.5% average lowering in frequency can be ascribed solely to the softening of the cantilever or whether stress relaxation is also considerable. Therefore, we removed the layer stack from the model as well, and repeated the calculation with the average Si Young's modulus value of $E_{Si} = 149.5$ GPa. As shown in the last two columns of Table 1, the agreement between the measured and calculated resonance frequencies for bare Si cantilevers is excellent. The obtained root-mean-square deviation of 5.8 Hz is remarkably low for such a simplified analytical model. It has to be noted that no fitting parameter was used for the calculation, only the measured geometrical and literature material parameters were used. Since the agreement was also good for the coated cantilevers, and no systematic shift was observed in the resonance frequencies, we can conclude that the effect of the AlN stress on the resonance frequency, if there is any, is negligible in these cantilevers.

3.3. Resonant Modes

As it was found by the detailed COMSOL analysis, the listed eigenfrequencies do not correspond to a pure vertical oscillation of the seismic mass (bending mode) but are also accompanied with a torsional vibration. As shown in Figure 7a, the displacement of the beam is not simply increasing from the clamping point to the center, but also superimposed with a periodic fluctuation. The off-axis wobbling motion of the mass was also confirmed by a stroboscopic observation of the resonating cantilever under optical microscope (Figure 7b).

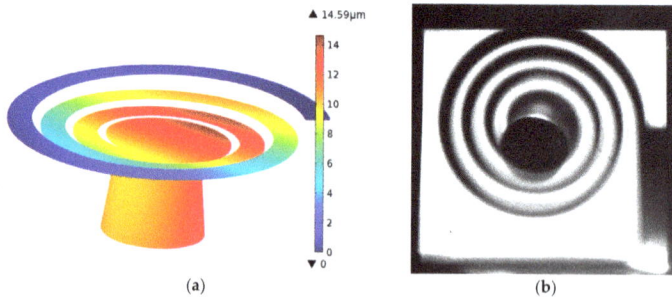

(a)   (b)

Figure 7. (**a**) Deflection distribution along the spiral cantilever at a driving acceleration of 1 g calculated by finite element analysis; (**b**) stroboscopic snapshot taken from the backside during shaking of the microspiral under optical microscope. The live slow-motion video clearly showed the off-axis wobbling movement of the seismic mass.

In accordance with these findings, the first bending mode frequencies calculated by a simplified straight rectangular model (Equation (2)) using the same length, width (Table 1), and thickness parameters are significantly lower than the corresponding experimental values shown in Figure 6. Nevertheless, by plotting these points against each other, a clear linear correlation is observed ($f_{exp} = 150 + 5.94 \cdot f_{bend}$), which can be used as a simple guideline for further chip designs (Figure 8). Repeated vibrational analyses performed in the range of 20–200 Hz revealed the presence of lower frequency resonances, presumably corresponding to these bending modes. However, they produce significantly lower voltage signals, which are hardly detectable in the background electrical noise.

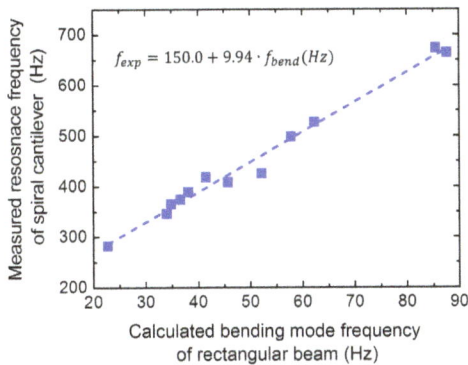

Figure 8. Experimental resonance frequencies for each cantilever as a function of the calculated first bending mode frequency using a simplified rectangular beam model.

3.4. Applicability in FICI

In the following, we will discuss the applicability of the piezo MEMS spirals in next generation FICI from the viewpoint of the tonotopy and the output voltage. The test arrays showed clear frequency selectivity in the 281–672 Hz range, which corresponds to the lowest part of the voice frequency regime. The high Q-factors, up to 254, enable high vibration sensitivity and frequency selectivity. By applying a shorter cantilever, it can be easily extended to the whole range up to 3400 Hz, or even to 20 kHz at the same or at a smaller device footprint. The ultralow frequency range down to the hearing limit of 20 Hz is more challenging, but seems to be also feasible for a 2 mm × 2 mm cantilever, assuming that we can enhance the torsion free vertical oscillating mode e.g., by a double beam sandwich structure [12]. One critical issue, which can cause false tones for a CI patient, is an unwanted overtone. With the demonstrated design, its exclusion was fairly successful in the 20–1200 Hz frequency range, since only very small second peaks are visible in Figure 6, and the ultralow frequency resonances were suppressed. However, the cantilever response in the upper part of the human audible range (1.2–20 kHz) was not investigated in the present study.

The obtained output voltage signals (3–10 mV) are higher than the ones reported by Jang et al. [11], therefore, it must be sufficient to excite hearing nerves through an implantable amplifier using a similar system. However, in this study, the excitation was not done directly by sound pressure, but by controlled acceleration of the cantilever frame. Moreover, its magnitude is much higher than the typical values for human ossicles. The displacement of the stapes footplate is about 30 nm at 500 Hz, and at a sound pressure level of 90 dB stimulus [17], which corresponds to an acceleration of 0.30 m/s². Assuming a linear relationship between the output voltage and the acceleration, the corresponding Channel 9 oscillator would result in an open circuit voltage of 100 μV, which is comparable to the magnitude which was detected and amplified by Jang et al. [11]. On the other hand, the ultimate solution of a FICI is a self-powered artificial basilar membrane which generates sufficient electronic power for the direct excitement of the hearing nerves through an implanted multielectrode. However, it requires further optimization in biocompatible piezoelectric materials, in CI electrode stimulation efficiency, and in implantation design.

4. Conclusions

In summary, we demonstrated a one arm spiral cantilever design and fully CMOS compatible fabrication procedure using biocompatible AlN piezoelectric layer and 3D micromachining. As it was confirmed by surface optical microscope and vibration tests, the AlN coated cantilevers are free of stress, which is essential for highly demanding human implants. The fabricated piezoelectric spiral array with Si tip mass exhibited a clear tonotopy in the lower audible frequency range (281–672 Hz).

The obtained high Q-factors (117–254) ensure high frequency selectivity for the targeted hearing sensor. To the analytical model of Karami et al. [23], we added extra terms to represent the seismic mass at the end of the spiral shape beam. The calculated natural resonance frequencies showed excellent agreement with the experimental values for coating free cantilevers (RMS = 5.8 Hz). The same method was also successfully used on AlN coated cantilevers, using a composite Young's modulus value for the layer stack. Besides, we also obtained a simple fit equation which helps to predict the main resonance frequencies of the spiral beams. All the manufactured cantilevers fit into a square of 2 mm × 2 mm, which satisfies the size constraint for middle ear implants. Thus, the feasibility of a frequency selective compact sensor array has been demonstrated that could provide a basis for further fully implantable cochlear implants.

Acknowledgments: The work (KoFAH, NVKP_16-1-2016-0018) was supported by the National Research, Development and Innovation Fund of the Hungarian Government. Clinical doctor, József Pytel and B&K engineer, Péter Szuhay are acknowledged for the fruitful discussions. The authors thank the contribution of Margit Payer for sample fabrication and Attila Nagy for wire bonding and chip packaging. The authors thank Viktor Kenderesi for the optical surface profilometer tests.

Author Contributions: The concept of the work was defined by Péter Udvardi and János Volk. Péter Udvardi carried out the design, finite element analysis, programmed the data acquisition card, and carried out the measurements. Zoltán Hajnal assisted in the design of the cantilevers and in COMSOL simulations. The fabrication of the device was carried out by András Straszner. János Radó built the testing setup, fabricated the chip holder for the shaker and assisted in the measurement. Michael Schneider and Ulrich Schmid supported the research with AlN layers during the optimization of the fabrication procedure. The final AlN layer described in the manuscript was deposited by János Ferencz. The idea of mounting the sensor on one of the ossicles is originated from Péter Révész. Saeedeh Soleimani assisted in the stress analysis of AlN. János Volk wrote the paper and coordinated the work.

Conflicts of Interest: The authors declare no conflict of interest.

References

1. Wilson, B.S.; Dorman, M.F. Cochlear implants: Current designs and future possibilities. *J. Rehabil. Res. Dev.* **2008**, *45*, 695–730. [CrossRef] [PubMed]
2. Cohen, N. The totally implantable cochlear implant. *Ear Hear.* **2007**, *28*. [CrossRef] [PubMed]
3. Zhou, G.; Bintz, L.; Anderson, D.Z.; Bright, K.E. A life-sized physical model of the human cochlea with optical holographic readout. *J. Acoust. Soc. Am.* **1993**, *93*, 1516–1523. [CrossRef] [PubMed]
4. Wittbrodt, M.J.; Steele, C.R.; Puria, S. Developing a physical model of the human cochlea using microfabrication methods. *Audiol. Neurotol.* **2006**, *11*, 104–112. [CrossRef] [PubMed]
5. White, R.D.; Grosh, K. Microengineered hydromechanical cochlear model. *Proc. Natl. Acad. Sci. USA* **2005**, *102*, 1296–1301. [CrossRef] [PubMed]
6. Shintaku, H.; Nakagawa, T.; Kitagawa, D.; Tanujaya, H.; Kawano, S.; Ito, J. Development of piezoelectric acoustic sensor with frequency selectivity for artificial cochlea. *Sens. Actuators Phys.* **2010**, *158*, 183–192. [CrossRef]
7. Inaoka, T.; Shintaku, H.; Nakagawa, T.; Kawano, S.; Ogita, H.; Sakamoto, T.; Hamanishi, S.; Wada, H.; Ito, J. Piezoelectric materials mimic the function of the cochlear sensory epithelium. *Proc. Natl. Acad. Sci. USA* **2011**, *108*, 18390–18395. [CrossRef] [PubMed]
8. Harada, M.; Ikeuchi, N.; Fukui, S.; Ando, S. Fish-bone-structured acoustic sensor toward silicon cochlear systems. In *Micromachined Devices and Components IV*; SPIE: Bellingham, WA, USA, 1998; Volume 3514. [CrossRef]
9. Xu, T.; Bachman, M.; Zeng, F.-G.; Li, G.-P. Polymeric micro-cantilever array for auditory front-end processing. *Sens. Actuators Phys.* **2004**, *114*, 176–182. [CrossRef]
10. Jang, J.; Kim, S.; Sly, D.J.; O'leary, S.J.; Choi, H. MEMS piezoelectric artificial basilar membrane with passive frequency selectivity for short pulse width signal modulation. *Sens. Actuators Phys.* **2013**, *203*, 6–10. [CrossRef]
11. Jang, J.; Lee, J.; Woo, S.; Sly, D.J.; Campbell, L.J.; Cho, J.-H.; O'Leary, S.J.; Park, M.-H.; Han, S.; Choi, J.-W.; et al. A microelectromechanical system artificial basilar membrane based on a piezoelectric cantilever array and its characterization using an animal model. *Sci. Rep.* **2015**, *5*, 12447. [CrossRef] [PubMed]

12. Choi, W.J.; Jeon, Y.; Jeong, J.-H.; Sood, R.; Kim, S.G. Energy harvesting MEMS device based on thin film piezoelectric cantilevers. *J. Electroceram.* **2006**, *17*, 543–548. [CrossRef]
13. Zhang, L.; Lu, J.; Takei, R.; Makimoto, N.; Itoh, T.; Kobayashi, T. S-shape spring sensor: Sensing specific low-frequency vibration by energy harvesting. *Rev. Sci. Instrum.* **2016**, *87*, 085005. [CrossRef] [PubMed]
14. Lu, J.; Zhang, L.; Yamashita, T.; Takei, R.; Makimoto, N.; Kobayashi, T. A silicon disk with sandwiched piezoelectric springs for ultra-low frequency energy harvesting. *J. Phys. Conf. Ser.* **2015**, *660*, 012093. [CrossRef]
15. Jackson, N.; Keeney, L.; Mathewson, A. Flexible-CMOS and biocompatible piezoelectric AlN material for MEMS applications. *Smart Mater. Struct.* **2013**, *22*, 115033. [CrossRef]
16. Heidrich, N.; Knöbber, F.; Sah, R.E.; Pletschen, W.; Hampl, S.; Cimalla, V.; Lebedev, V. Biocompatible AlN-based piezo energy harvesters for implants. In Proceedings of the 2011 16th International Solid-State Sensors, Actuators and Microsystems Conference, Beijing, China, 5–9 June 2011; pp. 1642–1644.
17. Gan, R.Z.; Dai, C.; Wang, X.; Nakmali, D.; Wood, M.W. A totally implantable hearing system—Design and function characterization in 3D computational model and temporal bones. *Hear. Res.* **2010**, *263*, 138–144. [CrossRef] [PubMed]
18. Beker, L.; Zorlu, Ö.; Göksu, N.; Külah, H. Stimulating auditory nerve with MEMS harvesters for fully implantable and self-powered cochlear implants. In Proceedings of the 2013 Transducers & Eurosensors XXVII: The 17th International Conference on Solid-State Sensors, Actuators and Microsystems (TRANSDUCERS & EUROSENSORS XXVII), Barcelona, Spain, 16–20 June 2013; pp. 1663–1666.
19. Ansari, M.Z.; Cho, C. Deflection, Frequency, and Stress Characteristics of Rectangular, Triangular, and Step Profile Microcantilevers for Biosensors. *Sensors* **2009**, *9*. [CrossRef] [PubMed]
20. Sharpe, W.N.; Yuan, B.; Vaidyanathan, R.; Edwards, R.L. Measurements of Young's modulus, Poisson's ratio, and tensile strength of polysilicon. In Proceedings of the IEEE Tenth Annual International Workshop on Micro Electro Mechanical Systems. An Investigation of Micro Structures, Sensors, Actuators, Machines and Robots, Nagoya, Japan, 26–30 January 1997; pp. 424–429.
21. Stokey, W.F. Vibration of systems having distributed mass and elasticity. In *Harris' Shock and Vibration Handbook*, 5th ed.; McGraw-Hill Education: New York City, NY, USA, 2002; p. 7.5.
22. Sooriakumar, K.; Chan, W.; Savage, T.S.; Fugate, C. A Comparative Study of Wet vs. Dry Isotropic Etch to Strengthen Silicon Micromachined Pressure Sensor. In *A Comparative Study of Wet vs. Dry Isotropic Etch to Strengthen Silicon Micromachined Pressure Sensor*; Electrochemical Society: Pennington, NJ, USA, 1995; pp. 259–265; ISBN 978-1-56677-123-8.
23. Karami, M.A.; Yardimoglu, B.; Inman, D. Coupled Out of Plane Vibrations of Spiral Beams. In Proceedings of the 50th AIAA/ASME/ASCE/AHS/ASC Structures, Structural Dynamics, and Materials Conference, Palm Springs, CA, USA, 4–7 May 2009.
24. Shackelford, J.F.; Alexander, W. Mechanical Properties of Materials. In *CRC Materials Science and Engineering Handbook*, 3rd ed.; CRC Press: Boca Raton, FL, USA, 2000; Table 407; ISBN 978-0-8493-2696-7.
25. Timoshenko, S.; Young, D.H. *Elements of Strength of Materials*, 4th ed.; Affiliated East-West Press: Delhi, India, 1962.
26. Hopcroft, M.A.; Nix, W.D.; Kenny, T.W. What is the Young's Modulus of Silicon? *J. Microelectromech. Syst.* **2010**, *19*, 229–238. [CrossRef]
27. Zhong, H.; Xiao, Z.; Jiao, X.; Yang, J.; Wang, H.; Zhang, R.; Shi, Y. Residual stress of AlN films RF sputter deposited on Si(111) substrate. *J. Mater. Sci. Mater. Electron.* **2012**, *23*, 2216–2220. [CrossRef]
28. Dubois, M.-A.; Muralt, P. Stress and piezoelectric properties of aluminum nitride thin films deposited onto metal electrodes by pulsed direct current reactive sputtering. *J. Appl. Phys.* **2001**, *89*, 6389–6395. [CrossRef]
29. Pobedinskas, P.; Bolsée, J.-C.; Dexters, W.; Ruttens, B.; Mortet, V.; D'Haen, J.; Manca, J.V.; Haenen, K. Thickness dependent residual stress in sputtered AlN thin films. *Thin Solid Films* **2012**, *522*, 180–185. [CrossRef]

micromachines

MDPI

Article

Design, Characterization and Sensitivity Analysis of a Piezoelectric Ceramic/Metal Composite Transducer

Muhammad bin Mansoor *, Sören Köble, Tin Wang Wong, Peter Woias and Frank Goldschmidtböing

Laboratory for the Design of Microsystems, Department of Microsystems Engineering (IMTEK), University of Freiburg, 79110 Freiburg, Germany; soeren.koeble@neptun.uni-freiburg.de (S.K.); wth1990@yahoo.com.hk (T.W.W.); woias@imtek.de (P.W.); frank.goldschmidtboeing@imtek.uni-freiburg.de (F.G.)
* Correspondence: mansoor@imtek.uni-freiburg.de; Tel.: +49-761-2036-7494

Received: 28 June 2017; Accepted: 31 August 2017; Published: 5 September 2017

Abstract: This article presents experimental characterization and numerical simulation techniques used to create large amplitude and high frequency surface waves with the help of a metal/ceramic composite transducer array. Four piezoelectric bimorph transducers are cascaded and operated in a nonlinear regime, creating broad band resonant vibrations. The used metallic plate itself resembles a movable wall which can align perfectly with an airfoil surface. A phase-shifted operation of the actuators results in local displacements that generate a surface wave in the boundary layer for an active turbulence control application. The primary focus of this article is actuator design and a systematic parameter variation experiment which helped optimize its nonlinear dynamics. Finite Element Model (FEM) simulations were performed for different design variants, with a primary focus in particular on the minimization of bending stress seen directly on the piezo elements while achieving the highest possible deflection of the vibrating metallic plate. Large output force and a small yield stress (leading to a relatively small output stoke) are characteristics intrinsic to the stiff piezo-ceramics. Optimized piezo thickness and its spatial distribution on the bending surface resulted in an efficient stress management within the bimorph design. Thus, our proposed resonant transduction array achieved surface vibrations with a maximum peak-to-peak amplitude of 500 μm in a frequency range around 1200 Hz.

Keywords: piezoelectric transducers; large stroke/high frequency vibrations; nonlinear duffing oscillator

1. Introduction

Recent progress in the field of aerospace engineering calls for new strategies to be developed aiming to influence the skin-friction drag. Modern research in aviation is focused around the central idea of influencing the aerodynamic flow behavior of the boundary layer near flight-relevant Reynolds number. Possible concepts are the maintenance of laminar flow, delaying the transition from laminar-to-turbulent flow, or influencing the turbulent boundary flow itself. Gad-el-Hak summarized the use of Micro-Electro-Mechanical-Systems (MEMS) actuators for active flow control applications [1]. Kline et al. proposed the most widely recognized idea of a near-wall autonomous and regenerative turbulence cycle, where the formation and interaction of local velocity fluctuations and coherent vortex structures takes place [2,3]. Hutschins et al. argued that at higher Reynolds numbers, the large-scale motion in the outer turbulent boundary layer can have a considerable effect on the near-wall turbulent cycles [4]. The hairpin structures align coherently in groups to form long packets, and packets align coherently to form very large-scale motions (VLSMs). Hence, the main objective of this industry/academia joint project was to research the area of turbulent flow control and identify suitable possibilities to influence the large-scale structures, resulting in reduced frictional drag.

Adrian et al. argued that large amplitude vibrations at frequencies in the region of 1–10 kHz can significantly influence the formation of large-scale coherent structures in the outer turbulent boundary layer [5]. Numerical simulations have also predicted that turbulence can be suppressed by the introduction of a transverse surface wave, leading to shear stress reduction [6]. Laadhari et al. have shown experimentally that turbulence could be suppressed by a spanwise oscillation of a wall section [7]. They produced sinusoidal oscillations (amplitude = 25 mm at frequencies in the region of 2–10 Hz) in a rectangular plate (1 m long, 0.7 m wide, and 10 mm thick) with a crank-shaft system. Roggenkamp et al. utilized an electromagnetic actuation mechanism to create a span-wise transversal surface wave, investigating vibration amplitudes of 250 to 375 μm at a frequency of 81 Hz [8]. Due to high input power density, high working frequency, and extremely high stiffness, piezoelectric actuators find themselves constantly in use for flow control applications. Warsop et al. developed a MEMS-based pulsed jet actuator operated by a piezoelectric cantilever for a flow-separation control experiment [9]. They produced air jets with a maximum speed of 300 m/s at frequencies up to 500 Hz through an orifice, having a diameter of 200 μm, modulated by the operation of a piezoelectric micro valve at 90 V. Haller et al. presented a smart array comprising a thin silicon membrane as a moveable wall with a closed surface for active cancellation of flow instabilities [10]. They utilized uni-morph and cymbal actuators as the excitation source, and fabricated their design with the piezo–polymer-composite technology, creating a surface wave of about 125 μm amplitude having a resonance frequency near 220 Hz. Hence, here we present a metal–ceramic-composite modular array achieving much larger vibrational amplitudes at comparatively higher frequencies. The metallic plate itself becomes the moveable wall in the airfoil surface and creates transverse (either traveling or stationary) surface vibrations within the turbulent boundary layer. The presented design shows system dynamics a factor of 20 faster as compared to what is commonly found in the literature.

2. Design Rationale

The actuator design concept entails the desired challenges from the realms of aerodynamics and continuum mechanics. Large amplitude surface vibrations (approximately 1 mm) at high frequencies (approximately 1–10 kHz) are required to align perfectly to an airfoil surface. Moreover, spatial limitations of the airfoil model restrict the maximum height of the transduction array to less than 25 mm. Such high dynamic motion becomes a challenging task once the required input power is considered. Extremely high strokes at high frequencies lead to extremely high accelerations, and therefore huge restoring forces on the moving part of the actuator. Hence choosing a very small dynamic mass—a 150 μm thin brass plate (20×90 mm^2, $\rho = 8440$ kg/m^3, $E = 110$ GPa) which would later become a part of the air-foil surface itself—is vital for an efficient actuator design. Surface geometry is designed to be rectangular, which provides an efficient way to cascade the actuators and form a modular transduction array. The actuation surface dimensions are chosen so as to create a traveling wave with a wavelength between 17 and 46 mm, actuating a $120 \times 100 = 12{,}000$ mm^2 surface area [11]. In order to ensure smooth alignment and avoid external edges in the boundary layer, the brass plate is glued along all edges with the help of two-component epoxy glue to an aluminum frame.

The basic actuation principle of the piezoelectric ceramic/metal composite actuator is the "bimorph-effect" (see Figure 1). Essentially, the design consists of two identical piezo elements (the active layer) symmetrically glued to a fully clamped rectangular brass plate (the passive layer). Application of a large external electric field to the piezo elements elongates (or contracts) the active layer while the passive layer tries to retain its original dimension, hence creating an inhomogeneous stress distribution in the ceramic/metal interface. This inhomogeneous stress distribution—resultant of the inherent piezoelectric electro-mechanical coupling—creates a bending moment, leading to an out-of-plane deflection of the plate.

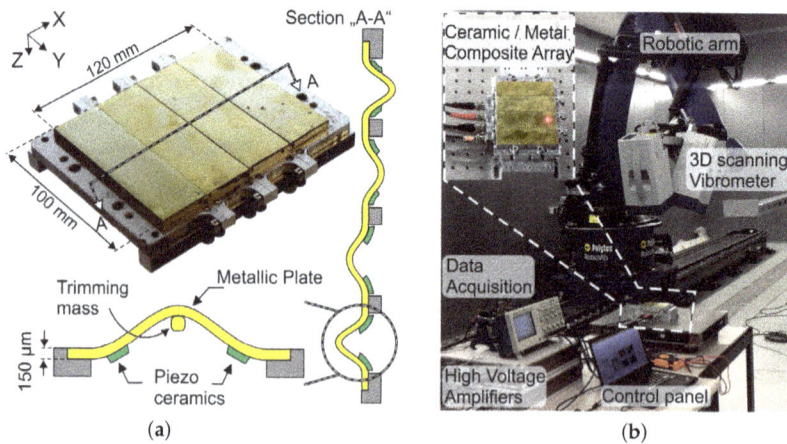

Figure 1. (**a**) 3D and cross-sectional view of the ceramic/metal composite array explaining its actuation principle. (**b**) Measurement setup showing the 3D PSV 500 scanning vibrometer (Polytec GmbH, Waldbronn, Germany) with a robotic arm measuring the entire actuation surface.

The phenomenon of resonance offers an energy-efficient operational window where energy required for the acceleration is already stored as the elastic potential energy in the system. Under steady-state conditions, only the damping forces need to be overcome by the exciting piezoelectric coupling force. Hence, the piezo elements are excited near the resonance frequency of the brass plate in order to vibrate the bimorph assembly into resonance. A complete analysis on the choice of actuation principle was also published elsewhere [12,13]. As the center displacement of the plate becomes larger than its thickness, in-plane stresses lead to a non-linear restoring force. This geometrically-induced stretching of the mid-plane—commonly modeled as an additional nonlinear spring—allows the actuator to achieve relatively higher strokes in a much broader range of frequencies compared to a linear oscillator with comparable quality factor.

3. Finite Element Modeling

Timoshenko has proposed mathematical expressions that can be utilized as "test functions" to analytically calculate the bending surface of a fully clamped rectangular plate [14]. However, symmetric application of the piezo elements to the clamped rectangular plate renders the task non-trivial. Now, a finite set of test functions are required both in x and y directions to accurately predict the bending surface while estimating the Ritz energy functions (similar to the derivations carried out for piezoelectric compound structures [15,16]).

Nevertheless, for such complicated geometries, finite element modeling (FEM) provides the most efficient solution. 3D numerical simulation, conducted using COMSOL Multiphysics 5.1 (COMSOL Inc., Stockholm, Sweden), helped to verify the experimentally obtained results. Here, an optimized device geometry together with the material data provided by the manufacturers (see Table 1) was implemented in a fully coupled electro-mechanical MEMS module for piezoelectric devices. The geometry was built using COMSOL's in-built Computer Aided Design (CAD) kernel, and the mesh was refined for free tetrahedral geometry with a maximum element size of 100 μm and at the contact boundaries with a maximum element size of 50 μm (see Figure 2). A "fixed-constraint" mechanical boundary condition was used in the solid-mechanics module to create an all-clamped rectangular geometry. An "electric-potential" terminal was used in the coupled electrostatics module generating 1 kV/mm electric field on the piezo-strips. Large signal contributions, which are intrinsic to the nonlinear piezoelectric behavior at high electric field strengths, are also externally added to the numerical model. A fully coupled stationary solver was implemented to solve 9.3×10^5 number of degrees of freedom at a relative tolerance of 0.001. The used finite element model allows a geometric parameter sweep to evaluate different device geometries, and numerical simulation results together with the experimentally measured values are presented in the next sections.

Table 1. Material constants for PZT-5K4 (Morgan Advanced Materials, Ohio, OH, USA) provided by the manufacturer: Morgan Advanced Materials (Ohio, OH, USA).

$\rho\ [\frac{kg}{m^3}]$	ν	Dielectric Constants		Compliance's $\times 10^{-12}[\frac{m^2}{N}]$		Charge Constants		$\times 10^{-12}[\frac{m}{V}]$
8300	0.31	$\epsilon_{33}^T = 7066$	$\epsilon_{11}^T = 6129$	$s_{33}^E = 20.03$	$s_{11}^E = 15.55$	$d_{33} = 926$	$d_{31} = -407$	$d_{15} = 950$

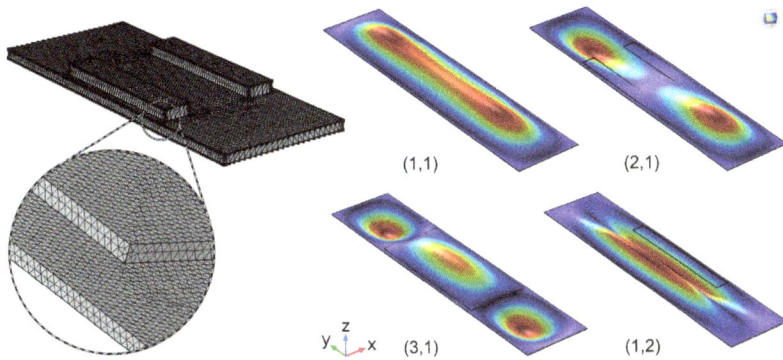

Figure 2. COMSOL Multiphysics implementation of the design with a tetrahedral element mesh and the expected vibration modes of the actuators.

4. Experimental Characterization

A systematic parametric analysis of the actuator's design has helped to optimize its performance near the desired frequency of operation. The most influential parameters—considering a large output amplitude at relatively higher frequencies—are (i) the thickness of piezo elements (h_p) in comparison to brass plate which defines the position of the neutral plane on the cross-section and (ii) the spatial distribution of piezo elements on vibrating composite (characterized as "ΔB"—see Figure 6). A detailed analysis is presented in the next sections highlighting the impact of these optimization parameters on the static and dynamic response of the actuator.

4.1. Test Setup

The quasi-static center displacement of the composite actuator up to a maximum of 1 kV/mm applied DC electric field was measured with the help of a 2 mm laser distance sensor (AWL 7/2, Welotech GmbH, Laer, Germany). For dynamic characterization, the piezo elements were excited near the first natural frequency of the brass plate with a sinusoidal signal sweep generated using a function generator. For high electric field actuation, a high-voltage amplifier was used (PZD 300 A, TREK INC., New York, NY, USA). Nonlinear resonant vibrations were measured with a 3D laser vibrometer (PSV 500, Polytec GmbH, Waldbronn, Germany) mounted on top of a robotic arm. The data were recorded and processed in a labVIEW program via a PC-interface (NI-cDAQ 9178, NI9263 and NI9223 (National Instruments, Austin, TX, USA)). The test setup is shown in Figure 3.

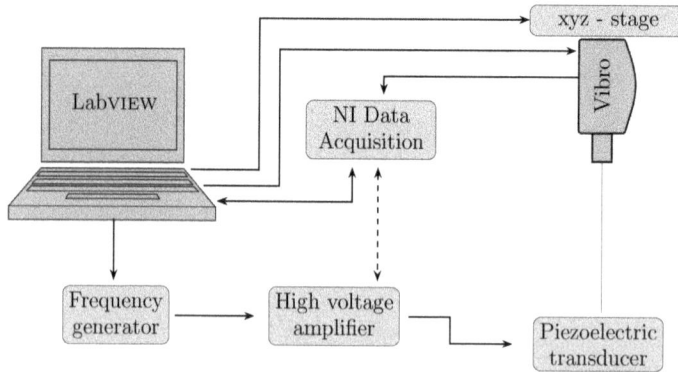

Figure 3. Schematic showing the dynamic experimental measurement setup.

4.2. Static Response

Although the actuator is designed to be used in resonance near the natural frequency of the brass plate, static analysis provided vital information which helped to optimize the device geometry. The first quadrant of the so called "butterfly-curve"—associated with intrinsic strain–electric field (S-E) hysteresis in the piezo elements—is shown in Figure 4. The measured center displacement of the plate (corresponding to the lateral strain generated by the piezo elements in the bimorph assembly) follows the so-called virgin curve for an increasing electric field during the first run. The measurement was repeated in a loop with successively decreasing maximum electric field value (E_{max}). Depending on the stress history of the transducer, a remnant strain S_{rem} leading to a remnant displacement at zero electric field was observed. Degree of hysteresis is an important figure of merit for piezoelectric transducers, characterizing the reproducibility of output stroke. Uchino [17] defined the degree of hysteresis in a transducer as:

$$Degree\ of\ Hysteresis\ \% = \frac{\Delta x}{X_{max}} \times 100 \tag{1}$$

where X_{max} is the displacement at maximum electric field E_{max}, and Δx is the difference in displacement for increasing and decreasing paths at half maximum of electric field, $E_{max}/2$. Degree of hysteresis in a piezoelectric transducer relates to the thermal and mechanical losses in the piezoelectric elements and/or the transducer design [18]. The main cause of these hysteretic losses in piezoelectric ceramics is the so-called domain reversal mechanism, which causes internal friction in the ceramic domains [19]. Whereas our proposed transducer design comprises an epoxy glue between the piezo elements and brass plate as well as the plate and the outer aluminum frame, making it the most probable cause of mechanical losses.

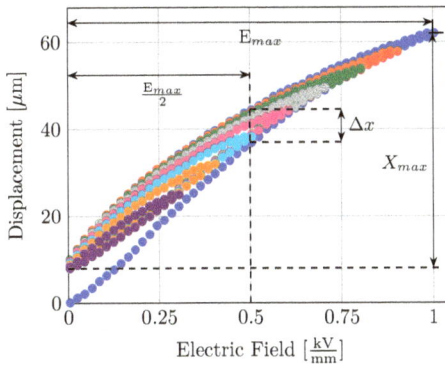

Figure 4. Uni-polar Strain–Electric field curve measured at successively decreasing maximum electric field strengths, showing the intrinsic hysteresis and remnant strain.

The basic goal of the parametric sensitivity analysis was to analyze the static and dynamic actuator characteristics when the active piezo elements were positioned near the clamped edges ($\Delta B = 1$ mm) as compared to the center of the brass plate ($\Delta B = 6$ mm—a common configuration used in many different piezoelectric actuation applications [20,21]). As ΔB increases in steps of 1 mm, the location of the piezo elements moves from the edges towards the center of the plate. The relative position of the piezo elements on the brass plate plays a significant role as we inspect the vibration characteristics of the composite corresponding to the first eigen mode. Six different variations of the parameter ΔB were fabricated (each actuator fabricated twice) and measured under the same experimental conditions.

Figure 5 shows numerically simulated bending lines of the quarter representation (shown in Figure 6) which also follow the experimentally measured trend. Relatively large variations of about 15 μm are seen along the $\Delta B = 1$ mm bending curve, which are attributed to a rough actuator surface. Numerically simulated bending lines at the mid-plane also show the characteristic "bimorph-bending" exactly at the position where the piezo-elements are located. Electric field sweeps measured at the mid point of the plate produce the previously mentioned S-E hysteresis curves. Note that these hysteresis curves are always plotted after 10 cycles in order to remove the zero point drift. The remnant strain S_{rem} and hence the remnant displacement also follows the familiar trend based on the relative vibrational amplitude seen directly at the piezo elements. Forward bending of the hysteresis curves is also observed—a common characteristic seen under high electric field operation. Displacement hysteresis among the design variants also show a dependence of the degree of hysteresis on the spatial distribution of piezo elements on the bending surface. Degree of hysteresis—as previously defined—is a characteristic of the width of the hysteresis loop and hence also corresponds to the dissipated energy density per loop [22]. As identical piezo-ceramics are utilized for all six configurations, this change in the dissipated energy density is solely attributed to the placement of piezo elements on the bending surface, and hence their effective strain amplitudes.

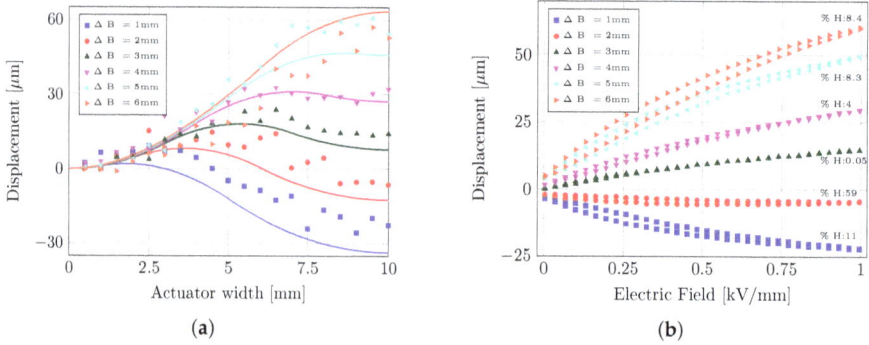

Figure 5. (a) Numerically simulated (solid lines) and experimentally measured (bold markers) bending lines of a quarter actuator model. (b) Hysteresis curves for a parametric variation of ΔB.

4.3. Dynamic Response

While the dimensions of the plate itself determine the frequency range of operation, the thickness of the piezo elements decides the location of the neutral axis on the ceramic/metal cross-section (see Figure 6). It is vital to keep the position of the neutral plane exactly at the contact interface between the piezo elements and the brass plate. This helps to avoid the active elements working against their own motion and hence to avoid internal losses. Figure 6 also shows the dynamic characterization results, where the typical hard Duffing-type nonlinearity is clearly observable in the frequency sweeps. Here the thickness of the piezo elements is reduced from 250 to 190 µm, which helps to avoid internal losses in the piezo elements. This change keeps the neutral plane at the contact interface, resulting in more than a factor of two increase in the peak-to-peak output vibrational amplitude. Apart from the piezo elements, there is another thin strip of brass attached exactly at the center of the vibrating plate. This thin strip adds mass to the vibrating assembly without influencing the bending surface, and hence acts a "frequency trimming mass". This mass helps to compensate the design tolerances and adjust the resonance frequency to the desired frequency of operation.

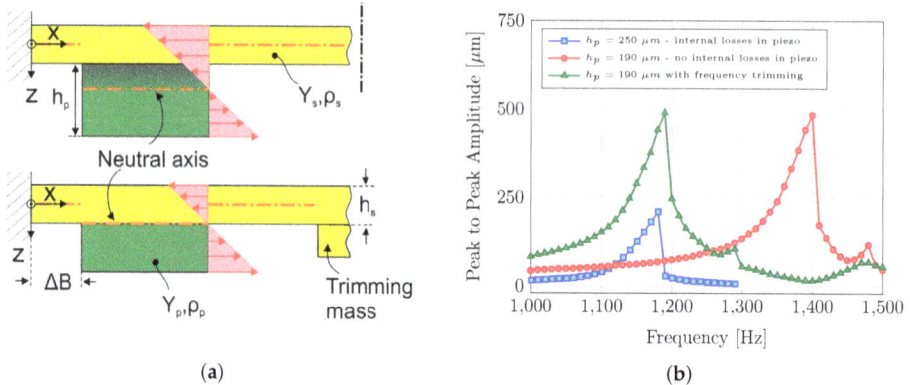

Figure 6. (a) Effect of piezo-thickness h_p on the non-homogenous strain distribution (solid arrows) and neutral axis position (dotted-dashed line). (b) Frequency sweeps near the first resonance frequency of the actuator. All results were measured at an electric field strength of 0.72 kV/mm.

Under quasi-static conditions, the actuation scheme with piezo elements directly placed in the middle of the plate (ΔB = 6 mm) favors the largest absolute bending, as the active elements are farthest away from the mechanically clamped edges (as seen in Figure 5). On the other hand, Figure 7 shows the dynamic response of the transducers under increasing electric field strengths for two extreme configurations (i.e., $\Delta B = 1$ mm compared to $\Delta B = 6$ mm). For high-frequency operations, the piezo elements at the center of the plate tend to break apart quite easily at relatively lower excitation field strengths. Transducers with piezo elements near the edges are relatively softer and have the first resonance vibration mode (1,1) near 1200 Hz. Here we also observe the third resonance peak (3,1), comparatively smaller than the first resonance peak. Depending on the position of the piezo elements, the effective thickness of the bending surface (and hence its stiffness) in comparison to the vibrating mass changes. This leads to an increased resonance frequency for the case $\Delta B = 6$ mm.

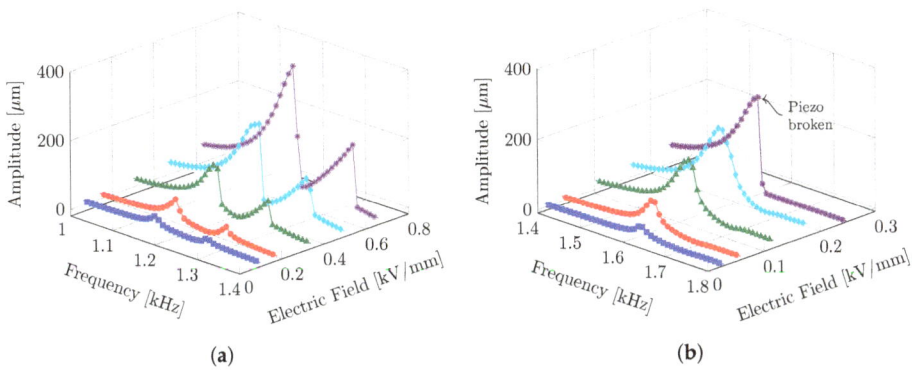

(a)   (b)

Figure 7. Frequency sweeps with increasing electric field strengths for transducer configurations with piezo elements near the edges ($\Delta B = 1$ mm—(**a**)) and at the center ($\Delta B = 6$ mm—(**b**)) of the plate.

5. Discussion

Considering the fact that the first eigen mode of vibration is similar to the static deflection of the transducer, we can draw comprehensive conclusions based on the static FEM simulation results. As piezo ceramics are known to be fragile towards tensile stresses, a "cost function" is defined in terms of the tensile stress σ_{22} seen along the active piezo elements and the maximum output absolute deflection of the composite at 1 kV/mm. For large amplitude vibration, minimizing the cost function would result in the strongest actuation with low stress levels in the active piezo elements. Figure 8 shows the numerically simulated cost function under quasi-static conditions for different transducer configurations along the x-axis cutline. Note that for clarity the cutline is only shown on a 2D simplification. At the ceramic boundaries along the x-axis cutline, a clear jump in the tensile stress distribution is observed. These jumps are accompanied by the irregular spikes, which are presumably a simulation anomaly arising due to a possible stress concentration at the ceramic boundaries (and hence a sudden jump in the neutral axis position). The maximum cost function value along the x-axis for $\Delta B = 6$ mm configuration is higher as compared to the $\Delta B = 1$ mm configuration. Here, the piezo elements themselves have to undergo the complete range of vibration as compared to transducers with piezo elements near the edges. Hence, the effective dynamic strain amplitude seen directly at the piezo elements is relatively larger when they are positioned at the middle of the plate. Thus, positioning the piezo elements near the edges allows them to withstand a much higher electric field strength while the composite vibrates in resonance. Due to the breakage of the brittle ceramics under resonant vibrations, positioning the piezo ceramics near the edges of the plate offers a considerable advantage for fast dynamic motion with large vibrational amplitudes.

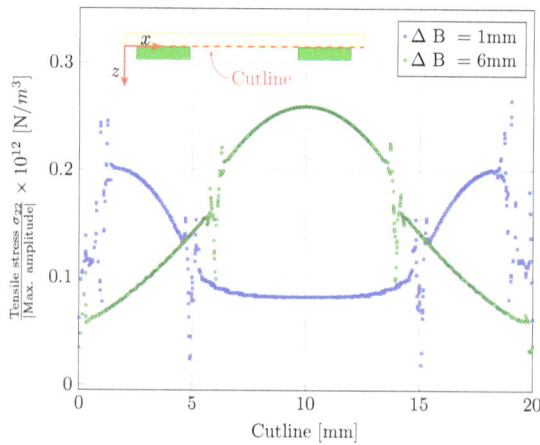

Figure 8. Numerically simulated cost-function (defined in terms of a ratio of tensile stress σ_{22} at the piezo to max. absolute output amplitude) across the x-axis cutline.

Finally, four such transducers are cascaded in a row to form an array capable of creating surface waves. Here the individual actuators are operated with a phase-shifted sinusoidal excitation signal at a frequency of 1200 Hz. Due to the above-mentioned nonlinear hardening behavior, a sweep from lower to higher frequencies is always required in order to vibrate the plate in the higher energy orbit. Figure 9 shows the 3D vibration surface of four such optimized transducers operated with and without a phase shift of 90° among each other. The vibration surface is measured with the 3D PSV vibrometers mounted on a robotic arm, as shown in Figure 1. The whole actuation surface is divided into a fine rectangular mesh of measurement points and recorded individually. As the phase information is also recorded with the vibrometer, a reconstruction of the measured data with the help of the automated PSV software from Polytec GmbH allows every mode to be visualized individually. Slight differences in the peak-to-peak values are attributed to different resonance frequencies of the transducers, which are minimized by frequency trimming but remain unavoidable.

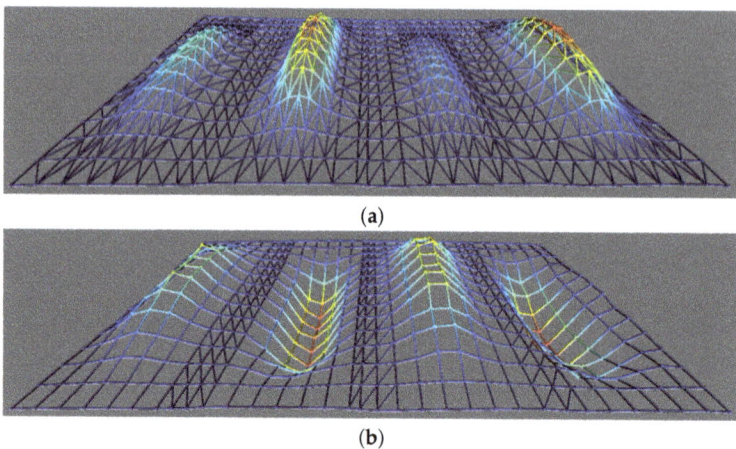

Figure 9. Surface vibrations of the whole array measured with the help of a 3D PSV vibrometer mounted on a robotic arm for in-phase (**a**) and 90° out-of-phase (**b**) actuation.

Micromachines **2017**, *8*, 271

6. Conclusions

Here we have presented a high energy density transduction scheme suitable for extremely fast system dynamics capable of creating either stationary or traveling surface waves in a turbulent boundary layer. Piezoelectric transducers are normally quite stiff and suitable for high-frequency operations. On the other hand, piezo ceramics are known to be particularly sensitive towards tensile loads. For large amplitude vibrations, the limiting factor for maximizing the stroke of the actuator is the mechanical failure of the brittle piezo ceramics, which break away at higher excitation field strengths. As observed in the parametric optimization results, ceramic/metal composite transducer shows the maximum absolute displacement under quasi-static conditions when the active elements are positioned near the middle of the plate. Here, the piezo elements themselves are positioned at the maximum displacement portion of the bending surface. On the contrary, positioning the piezo elements near the clamped edges have the advantage that the piezo-ceramics themselves do not have to undergo the maximum vibrational amplitude when the plate starts to vibrate in resonance. Therefore, the optimization criterion for a fast dynamic application is to have the smallest possible bending stress on the piezo elements while achieving maximum stroke of the vibrating plate.

Acknowledgments: The authors gratefully acknowledge the financial support from the Federal Ministry of Economics and Technology under the program LuFo V-1 "Ökoeffizientes Fliegen". The Project consortium also includes research groups from Airbus, Airbus Group, Deutsches Zentrum für Luft- und Raumfahrt (DLR) Göttingen, Universität der Bundeswehr München, Universität des Saarlands and Rheinisch-Westfälische Technische Hochschule (RWTH) Aachen.

Author Contributions: Muhammad bin Mansoor, Peter Woias and Frank Goldschmidtböing conceived and designed the experiments; Muhammad bin Mansoor and Sören Köble performed the experiments; Muhammad bin Mansoor and Frank Goldschmidtböing analyzed the data; Tin Wang Wong contributed materials/analysis tools; Muhammad bin Mansoor wrote the paper.

Conflicts of Interest: The authors declare no conflict of interest.

References

1. Gad-el Hak, M. Introduction to flow control. In *Flow Control*; Springer: Berlin, Germany, 1998; pp. 1–107.
2. Kline, S.J.; Reynolds, W.C.; Schraub, F.A.; Runstadler, P.W. The structure of turbulent boundary layers. *J. Fluid Mech.* **1967**, *30*, 741–773.
3. Jiménez, J.; Pinelli, A. The autonomous cycle of near-wall turbulence. *J. Fluid Mech.* **1999**, *389*, 335–359.
4. Hutchins, N.; Marusic, I. Large-scale influences in near-wall turbulence. *Philos. Trans. R. Soc. Lond. A* **2007**, *365*, 647–664.
5. Adrian, R.J. Hairpin vortex organization in wall turbulence. *Phys. Fluids* **2007**, *19*, 041301.
6. Du, Y.; Karniadakis, G.E. Suppressing wall turbulence by means of a transverse traveling wave. *Science* **2000**, *288*, 1230–1234.
7. Laadhari, F.; Skandaji, L.; Morel, R. Turbulence reduction in a boundary layer by a local spanwise oscillating surface. *Phys. Fluids* **1994**, *6*, 3218–3220.
8. Roggenkamp, D.; Jessen, W.; Li, W.; Klaas, M.; Schröder, W. Experimental investigation of turbulent boundary layers over transversal moving surfaces. *CEAS Aeronaut. J.* **2015**, *6*, 471–484.
9. Warsop, C.; Hucker, M.; Press, A.J.; Dawson, P. Pulsed air-jet actuators for flow separation control. *Flow Turbul. Combust.* **2007**, *78*, 255–281.
10. Haller, D.; Paetzold, A.; Losse, N.; Neiss, S.; Peltzer, I.; Nitsche, W.; King, R.; Woias, P. Piezo-polymer-composite unimorph actuators for active cancellation of flow instabilities across airfoils. *J. Intell. Mater. Syst. Struct.* **2011**, *22*, 461–474.
11. Pickel, C.; Sonnemann, D.; Ehlert, M.; Kähler, C.J. On the efficiency of active flow control with pneumatic jets at Mach numbers between 0.3 and 0.7. *Exp. Fluids* **2014**, *55*, 1683.
12. Bin Mansoor, M.; Koeble, S.; Woias, P.; Goldschmidtboeing, F. Design and optimization of nonlinear oscillators for drag reduction on airfoils. In Proceedings of the IEEE 29th International Conference on Micro Electro Mechanical Systems (MEMS), Shanghai, China, 24–28 January 2016; pp. 1129–1132.

13. Bin Mansoor, M.; Reuther, N.; Köble, S.; Gérard, M.; Steger, H.; Woias, P.; Goldschmidtböing, F. Parametric modeling and experimental characterization of a nonlinear resonant piezoelectric actuator designed for turbulence manipulation. *Sens. Actuators A Phys.* **2017**, *258*, 14–21.

14. Timoshenko, S.P.; Woinowsky-Krieger, S. *Theory of Plates and Shells*; McGraw-Hill: New York, NY, USA, 1959.

15. Hagood, N.W.; Chung, W.H.; Von Flotow, A. Modelling of piezoelectric actuator dynamics for active structural control. *J. Intell. Mater. Syst. Struct.* **1990**, *1*, 327–354.

16. Goldschmidtboeing, F.; Woias, P. Characterization of different beam shapes for piezoelectric energy harvesting. *J. Micromech. Microeng.* **2008**, *18*, 104013.

17. Uchino, K. *Piezoelectric Actuators and Ultrasonic Motors*; Springer Science & Business Media: Berlin, Germany, 1997; Volume 1.

18. Dogan, A.; Uchino, K.; Newnham, R.E. Composite piezoelectric transducer with truncated conical endcaps "Cymbal". *IEEE Trans. Ultrason. Ferroelectr. Freq. Control* **1997**, *44*, 597–605.

19. Jaffe, B. *Piezoelectric Ceramics*; Elsevier: Amsterdam, The Netherlands, 2012; Volume 3.

20. Doll, A.; Heinrichs, M.; Goldschmidtboeing, F.; Schrag, H.-J.; Hopt, U.T.; Woias, P. A high performance bidirectional micropump for a novel artificial sphincter system. *Sens. Actuators A Phys.* **2006**, *130*, 445–453.

21. Goldschmidtböing, F.; Doll, A.; Heinrichs, M.; Woias, P.; Schrag, H.-J.; Hopt, U.T. A generic analytical model for micro-diaphragm pumps with active valves. *J. Micromech. Microeng.* **2005**, *15*, 673–683.

22. Goldschmidtboeing, F.; Eichhorn, C.; Wischke, M.; Kroener, M.; Woias, P. The influence of ferroelastic hysteresis on mechanically excited PZT cantilever beams. In Proceedings of the 11th International Workshop on Micro and Nanotechnology for Power Generation and Energy Conversion Applications, Seoul, Korea, 15–18 November 2011; pp. 114–117.

micromachines

MDPI

Article

Parametric Analysis and Experimental Verification of a Hybrid Vibration Energy Harvester Combining Piezoelectric and Electromagnetic Mechanisms

Zhenlong Xu [1], Xiaobiao Shan [2], Hong Yang [3], Wen Wang [1] and Tao Xie [2,*]

[1] School of Mechanical Engineering, Hangzhou Dianzi University, Hangzhou 310018, China;
    xzl@hdu.edu.cn (Z.X.); wangwn@hdu.edu.cn (W.W.)
[2] School of Mechatronics Engineering, Harbin Institute of Technology, Harbin 150001, China;
    shanxiaobiao@hit.edu.cn
[3] College of Environmental and Resource Sciences, Zhejiang University, Hangzhou 310029, China;
    hyang_zju@163.com
*   Correspondence: xietao@hit.edu.cn; Tel.: +86-451-8641-7891

Received: 11 May 2017; Accepted: 15 June 2017; Published: 18 June 2017

Abstract: Considering coil inductance and the spatial distribution of the magnetic field, this paper developed an approximate distributed-parameter model of a hybrid energy harvester (HEH). The analytical solutions were compared with numerical solutions. The effects of load resistances, electromechanical coupling factors, mechanical damping ratio, coil parameters and size scale on performance were investigated. A meso-scale HEH prototype was fabricated, tested and compared with a stand-alone piezoelectric energy harvester (PEH) and a stand-alone electromagnetic energy harvester (EMEH). The peak output power is 2.93% and 142.18% higher than that of the stand-alone PEH and EMEH, respectively. Moreover, its bandwidth is 108%- and 122.7%-times that of the stand-alone PEH and EMEH, respectively. The experimental results agreed well with the theoretical values. It is indicated that the linearized electromagnetic coupling coefficient is more suitable for low-level excitation acceleration. Hybrid energy harvesting contributes to widening the frequency bandwidth and improving energy conversion efficiency. However, only when the piezoelectric coupling effect is weak or medium can the HEH generate more power than the single-mechanism energy harvester. Hybrid energy harvesting can improve output power even at the microelectromechanical systems (MEMS) scale. This study presents a more effective model for the performance evaluation and structure optimization of the HEH.

Keywords: hybrid energy harvester; piezoelectric; electromagnetic; approximate distributed-parameter model; parametric analysis

1. Introduction

Vibration energy harvesting technology, which converts the ambient vibration energy into electric energy, has drawn much attention in recent years. It is considered as a promising solution to power the low-power portable microelectronic devices and wireless sensor networks. The conventional conversion mechanisms comprise piezoelectric [1], electromagnetic [2], electrostatic [3] and magnetostrictive [4] types. It is a challenge for researchers to design the vibration energy harvester (VEH) with a broad operating frequency bandwidth and outstanding energy density. Because many reported VEHs are based on the principle of a linear single-degree-of-freedom (SDOF) system, only when the resonant frequencies match the excitation frequencies can they achieve the optimum generating performance. However, the ambient vibration frequency is broadband and random. To improve the performance of the VEH, hybrid energy harvesting technology combining piezoelectric and electromagnetic mechanisms is proposed. It shows an increasing trend and attracts more and more attention from scholars.

Generally, the hybrid energy harvester (HEH) is transformed from the piezoelectric energy harvester (PEH) by replacing the proof mass with a permanent magnet and adding an induction coil. The PEH generates electricity by means of the piezoelectric effect. The magnet is used to tune the resonant frequency and amplify the deformation of the piezoelectric element. Meanwhile, the relative motion between magnet and coil can induce electric current in the coil due to Faraday's law of electromagnetic induction. Both the magnet and induction coil are the components of the electromagnetic energy harvester (EMEH). Considering different vibration sources and application backgrounds, scholars conducted a series of structural designs for the HEH, such as wearable HEH [5], cantilever-type HEH [6,7], multimodal HEH [8–10], multi-frequency HEH [11], nonlinear HEH [12,13] and frequency up-converted HEH [14,15]. In a word, the beam structure is the most typical configuration of the reported HEH. With the development of microelectromechanical systems (MEMS), the dimension of the HEH shrinks from meso to micro size [16]. In order to evaluate the generating performance of micro-scale HEH, it is meaningful to investigate the scaling effects on the mechanical and electrical properties, which has not been reported up to now.

In addition, some scholars also make efforts to develop an effective theoretical model of the HEH to predict the generating performance and analyze the effects of parameters on the energy harvesting characteristics. Mostly, the HEH is simplified to be a spring-mass-damper system, and the lumped-parameter theoretical model [17] is established. However, many models [5–7,9,10] ignored the effects of piezoelectric and electromagnetic coupling coefficients on the effective stiffness of the system. Both Li et al. [18] and Shan et al. [19] ignored the influence of coil inductance. The approximate distributed-parameter model [12,20] derived from the energy method is another common model, which is more accurate than the lumped-parameter one. It takes into account the effects of mode shape, strain distribution, distributed mass and stiffness on the performance of the energy harvester. In the process of mathematical modeling, the electromagnetic coupling coefficient is mostly oversimplified and considered to be a constant. The effects of the nonlinear and linear electromagnetic coupling coefficients on the output power are not clear. Moreover, the effect of the coil inductance mostly is not taken into account. Therefore, it is disadvantageous for the performance evaluation and structure optimization of the HEH based on those theoretical models.

In this paper, a hybrid piezoelectric-electromagnet energy harvester is modeled, theoretically analyzed and experimentally tested. The objective of this paper is to develop an approximate distributed-parameter theoretical model of the HEH by considering the coil inductance and spatial distribution of magnetic field and to analyze comprehensively the effects of key parameters on the generating performance.

2. Modeling

Based on the previous research articles, a typical HEH with the cantilever-beam structure is designed, as depicted in Figure 1. It comprises a bimorph piezoelectric cantilever beam with a permanent magnet as the tip mass and a cylindrical induction coil attached on the frame. The piezoelectric beam is fixed on the base, whose oscillation is harmonic. The piezoceramic operates in the d_{31} mode. Both piezoceramic layers are assumed to be perfectly bonded to both sides of the substrate, respectively. They are oppositely poled in the thickness direction and connected in series. The conductive electrodes fully cover the top and bottom surfaces of the piezoceramic layers. The proof magnet is considered as a mass point without the rotary inertia. Under the small-amplitude oscillation, the Euler–Bernoulli beam theory is applicable to the piezoelectric beam. As a consequence, the rotary inertia and shear deformation of the beam are neglected. In addition, the coil and magnet are axially aligned.

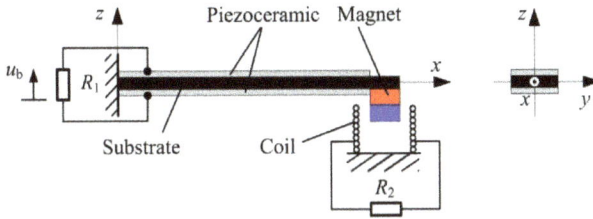

Figure 1. Schematic diagram of the hybrid energy harvester.

The electromechanical coupling model of the HEH can be derived by the energy method [21,22]. According to Hamilton's principle, the energies of the electromechanical system satisfy the following equation:

$$\int_{t_1}^{t_2} [\delta(T - U) + \delta W_{nc}] dt = 0 \tag{1}$$

where T, U and W_{nc} denote the kinetic energy, potential energy and virtual work of the system, respectively. δ is the variational operator.

When the HEH is subjected the transverse harmonic excitation along the z-axis, only the transverse displacement is considered. The kinetic energy T of the HEH is the sum of the translational kinetic energies of the cantilever beam and magnet, which is defined as:

$$T = \frac{1}{2}\int_{V_s} \rho_s \dot{u}^t \dot{u} dV_s + \frac{1}{2}\int_{V_p} \rho_p \dot{u}^t \dot{u} dV_p + \frac{1}{2}M_t \dot{u}(L)^t \dot{u}(L) \tag{2}$$

where V and ρ represent the volume and density, respectively. The subscripts "s" and "p" denote the substrate and piezoceramic, respectively. M_t is the mass of the magnet. L is the length of the beam. $u(x, t)$ is the displacement of the piezoelectric beam relative to the base at the axial position x and time t. Dots indicate the time derivatives. Therefore, $\dot{u}$ is the velocity. The superscript "t" denotes the transpose of the matrix.

The potential energy U consists of four terms: the elastic potential energy stored in the substrate and piezoceramic layers, electric energy stored in the piezoceramic layers and the magnetic co-energy stored in the induction coil. It is given by:

$$U = \frac{1}{2}\int_{V_s} S^t T dV_s + \frac{1}{2}\int_{V_p} S^t T dV_p + \frac{1}{2}\int_{V_p} E^t D dV_p + \frac{1}{2}L_c \dot{Q}^2 \tag{3}$$

where S, T, E and D are the strain, stress, electric field and electric displacement, respectively. Q is the charge passing through the coil. L_c is the inductance of the coil.

The virtual work W_{nc} applied to the HEH contains the mechanical work done by electromagnetic force, base excitation force, mechanical damping force and electrical work due to the charges. Hence, the virtual work can be expressed as:

$$\delta W_{nc} = -\delta u(L)^t M_t \ddot{u}_b - \int_{V_s} \delta u^t \rho_s \ddot{u}_b dV_s - \int_{V_p} \delta u^t \rho_p \ddot{u}_b dV_p - \int_0^L C \dot{u} \delta u dx$$
$$-\delta Q^t (R_c + R_2)\dot{Q} + \sum_{j=1}^{nq} \delta\phi_j q_j + \theta_e \dot{Q}\delta u(L) \tag{4}$$

where $u_b(t)$ is the base displacement and $\ddot{u}_b$ is the acceleration. C denotes the viscous damping, which can be experimentally determined by the logarithmic decrement method [23]. R_c and R_2 are the internal resistance of the coil and the external load resistance connected to the coil, respectively. $\phi_j = \phi(x_j, t)$ is the scalar electrical potential for each of the nq electrode pairs at position x_j. q_j is the charge extracted from the corresponding electrode pairs. θ_e is the electromagnetic coupling coefficient.

Due to the low frequency of the ambient excitation, the first mode is our research focus. Based on the small-signal constitutive equations of piezoelectricity [24], the relationship between the stress and strain of the substrate, the Rayleigh–Ritz approach and the Euler–Bernoulli beam assumption, the modal electromechanical coupling equations of the HEH are given by:

$$\begin{cases} M\ddot{r} + C\dot{r} + Kr - \theta_p v - \theta_{em} I_2 = -B_f \ddot{u}_b \\ \theta_p \dot{r} + C_p \dot{v} + v/R_1 = 0 \\ \theta_{em} \dot{r} + L_c \dot{I}_2 + (R_c + R_2) I_2 = 0 \end{cases} \tag{5}$$

where $v = R_1 \dot{q}$ is the voltage across the piezoelectric beam. R_1 is the external load resistance connected to the beam. I_2 is the current in the coil. Note that $u(x, t) = \psi_r(x)r(t)$, where $\psi_r(x)$ and $r(t)$ denote the mode shape and mechanical temporal coordinate, respectively. M, K, θ_p, θ_{em}, B_f and C_p are the effective mass, effective stiffness, piezoelectric coupling term, electromagnetic coupling term, forcing coefficient and capacitance, respectively. θ_{em} is equal to the product of θ_e and $\psi_r(L)$.

The calculation of the electromagnetic coupling coefficient θ_e is related to the accuracy of the mathematical model and the optimization of system parameters. In this paper, it is established based on the magnetic dipoles model, which has been reported in our previous study [11]. The result is expressed as:

$$\theta_e = -\frac{B_r V_m f_c N}{2A_c} \left[\ln \frac{R_i + \sqrt{R_i^2 + (z_2 - h_c)^2}}{R_o + \sqrt{R_o^2 + (z_2 - h_c)^2}} + \ln \frac{R_o + \sqrt{R_o^2 + z_2^2}}{R_i + \sqrt{R_i^2 + z_2^2}} \right.$$
$$\left. + \frac{R_o}{\sqrt{R_o^2 + (z_2 - h_c)^2}} - \frac{R_o}{\sqrt{R_o^2 + z_2^2}} - \frac{R_i}{\sqrt{R_i^2 + (z_2 - h_c)^2}} + \frac{R_i}{\sqrt{R_i^2 + z_2^2}} \right] \tag{6}$$

where B_r and V_m are the residual magnetic flux density and volume of the magnet, respectively. R_i, R_o, h_c, f_c and N are the inner radius, outer radius, height, fill factor and turns of the induction coil, respectively. $A_c = (R_o - R_i)h_c$ is the cross-section area of the coil. z_2 is the position coordinate of the magnet core relative to the coil. It can be seen that there is a nonlinear change of θ_e with the change of magnet position z_2. This model takes the spatial distribution of the magnetic field into account.

When the HEH is excited by the harmonic vibration, the beam-tip displacement relative to the base is $u(L)$. Therefore, z_2 can be expressed as $z_2 = u(L) + z_0$, where z_0 is the position coordinate of the magnet core at static balance. A Taylor series expansion of θ_e about z_0 is:

$$\theta_e = \sum_{i=0}^{\infty} \left. \frac{\partial^i \theta_e}{\partial z_2^i} \right|_{z_2 = z_0} \cdot \frac{u^i(L)}{i!} \tag{7}$$

For convenience, θ_e can be simplified to $\theta_e(z_0)$ in the linearized model under small signal excitation.

Defining a state vector $\mathbf{X} = \begin{bmatrix} X_1 & X_2 & X_3 & X_4 \end{bmatrix}^t = \begin{bmatrix} r & \dot{r} & v & I_2 \end{bmatrix}^t$ (t denotes the transpose of the vector here), Equation (5) can be written in the state space form as:

$$\dot{\mathbf{X}} = \begin{bmatrix} X_2 \\ -\frac{K}{M}X_1 - \frac{C}{M}X_2 + \frac{\theta_p}{M}X_3 + \frac{\theta_{em}}{M}X_4 - \frac{B_f}{M}\ddot{u}_b \\ -\frac{\theta_p}{C_p}X_2 - \frac{1}{C_p R_1}X_3 \\ -\frac{\theta_{em}}{L_c}X_2 - \frac{R_c + R_2}{L_c}X_4 \end{bmatrix} \tag{8}$$

The output average power delivered to the external loads R_1 and R_2 is respectively given as:

$$P_{np} = \frac{1}{T} \int_0^T v^2/R_1 dt \tag{9}$$

$$P_{\text{ne}} = \frac{1}{T}\int_0^T I_2^2 R_2 dt \tag{10}$$

where $T = 2\pi/\omega$ is the cycle of the base excitation. ω is the angular velocity.

The total output power of the HEH with nonlinear θ_e is:

$$P_1 = P_{\text{np}} + P_{\text{ne}} \tag{11}$$

When using the linearized model of θ_e, the analytical solutions of the amplitude for relative displacement, voltage across the load resistance R_1 and current through the coil can be derived from Equation (5), as shown below:

$$|r| = \frac{|\ddot{u}_b|B_f}{K}\cdot\left[\left(1-\Omega^2+\frac{\kappa_p^2\lambda_p^2\Omega^2}{1+\lambda_p^2\Omega^2}+\frac{\kappa_e^2\lambda_e^2\Omega^2}{1+\lambda_e^2\Omega^2}\right)^2+\left(2\zeta_m\Omega+\frac{\kappa_p^2\lambda_p\Omega}{1+\lambda_p^2\Omega^2}+\frac{\kappa_e^2\lambda_e\Omega}{1+\lambda_e^2\Omega^2}\right)^2\right]^{-1/2} \tag{12}$$

$$|v| = \frac{\theta_p}{C_p}\cdot\frac{\lambda_p\Omega}{\sqrt{1+\lambda_p^2\Omega^2}}\cdot|r| \tag{13}$$

$$|I_2| = \frac{\theta_{\text{em}}}{L_c}\cdot\frac{\lambda_e\Omega}{\sqrt{1+\lambda_e^2\Omega^2}}\cdot|r| \tag{14}$$

where $\Omega = \omega/\omega_1$ is the excitation frequency ratio. $\omega_1 = \sqrt{K/M}$ is the first undamped natural frequency of the piezoelectric beam. $\kappa_p^2 = \theta_p^2/(KC_p)$ and $\kappa_e^2 = \theta_{\text{em}}^2/(KL_c)$ are the effective electromechanical coupling coefficients for the PEH and EMEH, respectively. $\lambda_p = R_1C_p\omega_1$ and $\lambda_e = L_c\omega_1/(R_c+R_2)$ are the normalized load resistances connected to the PEH and EMEH, respectively. $\zeta_m = C/(2\sqrt{KM})$ is the mechanical damping ratio.

The instantaneous output power of the PEH is calculated as $P_p = v^2/R_1$, and that of EMEH is $P_e = I_2^2R_2$. Therefore, the total average output power of the HEH with linear θ_e is:

$$P_2 = \overline{P}_p + \overline{P}_e = \frac{r_{\text{max}}^2}{2}\left[\frac{K\omega_1\kappa_p^2\lambda_p\Omega^2}{1+\lambda_p^2\Omega^2}+\frac{K\omega_1\kappa_e^2\lambda_e\Omega^2}{1+\lambda_e^2\Omega^2}\cdot\left(1-\frac{\lambda_e}{\lambda_c}\right)\right] \tag{15}$$

where $\lambda_c = L_c\omega_1/R_c$ is the normalized internal resistance of the coil. r_{max} denotes the magnitude of the relative displacement.

The harvested vibration energy from the base excitation is ultimately dissipated due to the mechanical damping and electric resistances. We have to note that there is power loss due to the internal resistance of the coil. In general, this part is neglected by researchers [25,26]. The average power dissipated by the mechanical damping and internal resistance of the coil is respectively expressed as:

$$\overline{P}_d = \frac{\omega}{2\pi}\int_0^{2\pi/\omega}C\dot{r}^2 dt = \frac{1}{2}C\omega^2 r_{\text{max}}^2 \tag{16}$$

$$\overline{P}_c = \overline{P}_e\cdot\frac{R_c}{R_2} \tag{17}$$

As a result, the energy conversion efficiency is obtained by:

$$\eta = \frac{P_2}{P_2+\overline{P}_d+\overline{P}_c} \tag{18}$$

From Equation (12), it can be seen that the piezoelectric and electromagnetic coupling affect the effective stiffness and damping of the system. The effective damping of the HEH is:

$$C_{\text{eff}} = C + \frac{K\kappa_{\text{p}}^2\lambda_{\text{p}}}{\omega_1(1+\lambda_{\text{p}}^2\Omega^2)} + \frac{K\kappa_{\text{e}}^2\lambda_{\text{e}}}{\omega_1(1+\lambda_{\text{e}}^2\Omega^2)} = C + D_{\text{p}} + D_{\text{em}} \tag{19}$$

where D_{p} and D_{em} are respectively defined as the piezoelectric and electromagnetic damping. The power angular bandwidth [27] of the HEH can be derived by:

$$\Delta\omega = 2\zeta\omega_1 = 2(\zeta_{\text{m}} + \zeta_{\text{e}})\omega_1 = \frac{C + D_{\text{p}} + D_{\text{em}}}{M} \tag{20}$$

where ζ is the total damping ratio of the HEH. $\zeta_{\text{e}} = (D_{\text{p}} + D_{\text{em}})/(2\sqrt{KM})$ denotes the electrical damping ratio, which is the sum of the piezoelectric and electromagnetic damping ratios.

3. Fabrication and Parametric Analysis

In this section, numerical and analytical solutions of the output power of the HEH for different excitation frequency ratios and accelerations are obtained by using MATLAB software (R2012b, MathWorks Inc., Natick, MA, USA) and compared firstly, so as to analyze the effect of electromagnetic coupling coefficient. The analytical solutions are used to investigate the influences of electric load resistances, electromechanical coupling factors, mechanical damping ratio, coil parameters and size scale on the generating characteristics and dynamic responses of the HEH with the acceleration of the harmonic base excitation 2 m/s^2.

Firstly, a meso-scale HEH prototype based on the proposed structure was fabricated, as shown in Figure 2. The bimorph piezoelectric cantilever beam is made of phosphor bronze substrate and two lead zirconate titanate (PZT-5H) layers. The magnet material is NdFeB (N35). The induction coil is wound with copper wire. The geometric and material parameters of the HEH are listed in Table 1, where c_{11}^E, e_{31} and ε_{33}^S are experimentally obtained by using the impedance analyzer (Agilent 4294A). For the convenience of qualitative analysis, the magnetic core is located at the same height as the upper surface of the coil at the static state. The default of the mechanical damping ratio ζ_{m} is set to 0.02.

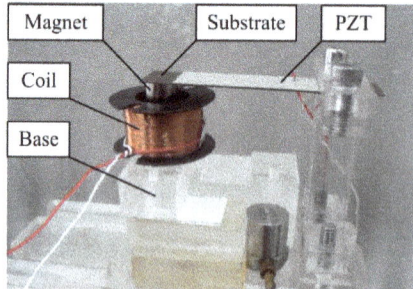

Figure 2. Prototype of the hybrid energy harvester (HEH).

Table 1. Geometric and material parameters of the HEH.

Parameter	Value
Substrate length × width × thickness (mm^3) $L \times b \times h_{\text{s}}$	62 × 20 × 0.26
Substrate density (kg/m^3) ρ_{s}	8920
Substrate Young's modulus (GPa) c_{s}	90
PZT length × width × thickness (mm^3) $L \times b \times h_{\text{p}}$	50 × 20 × 0.2
PZT density (kg/m^3) ρ_{p}	7386
PZT elastic stiffness (GPa) c_{11}^E	59.77
Piezoelectric stress constant (C/m^2) e_{31}	−13.74
Dielectric permittivity (nF/m) ε_{33}^S	38.62

Table 1. *Cont.*

Parameter	Value
Magnet radius × height (mm^2) $R \times h_{\mathrm{m}}$	6 × 10
Residual magnetic flux density (T) B_r	1.25
Magnet density (kg/m^3) ρ_{m}	7800
Coil turns N	2050
Coil inner radius × outer radius × height (mm^3) $R_{\mathrm{i}} \times R_{\mathrm{o}} \times h_{\mathrm{c}}$	11 × 13.15 × 17
Wire diameter (mm) D_{w}	0.123
Wire resistance per unit length (Ω/m) ρ_{c}	2.16

3.1. Comparison of Numerical and Analytical Solutions

Figure 3 shows the comparison of output power obtained from numerical and analytical solutions at 2-, 10-, 20- and 40-m/s^2 excitation acceleration. Note that P_{ap} and P_{ae} denote the analytical results generated from PEH and EMEH subsystems, respectively. The values of R_1 and R_2 are $1/(C_{\mathrm{p}}\omega_1)$ and R_{c}, respectively. Define ω_{r} as the resonant frequency of the HEH with load resistances and $\Omega_{\mathrm{r}} = \omega_{\mathrm{r}}/\omega_1$ as the normalized resonant frequency. From Figure 3, we can see that the resonant frequency ratio Ω_{r} of HEH is not affected by the excitation acceleration and remains unchanged. However, the difference between the output power of numerical and analytical results gradually increases with the increment of excitation acceleration. This is due to the effect of electromagnetic coupling coefficient θ_{e}. In the linearized model, θ_{e} is considered to be a constant, which is almost equal to the maximum value of nonlinear θ_{e} in the steady state. Consequently, the electromagnetic damping force induced by the linearized θ_{e} is much larger than that in the numerical model. The vibration response of piezoelectric beam is suppressed more obviously, which leads to the decrease of the output power from the PEH subsystem. Furthermore, the effect of vibration suppression is enhanced with the increasing of the excitation acceleration, resulting in the enlargement of the difference between numerical and analytical solutions. Although the output power generated from the EMEH subsystem is related with θ_{e}, the relative velocity between magnet and induction coil is another determining factor. The magnet is slowed down along with the vibration suppression. Therefore, the output power of the EMEH subsystem also decreases. It is indicated that the linearized model of θ_{e} is more suitable to predict the generating performance of the HEH in the low excitation acceleration level.

Table 2 lists the resonant frequency ratios and peak output power at different excitation accelerations obtained from numerical and analytical solutions. Figure 4 shows the deviation rates of analytical values relative to the numerical values. "h", "p" and "e" in the legend represent the HEH, PEH subsystem and EMEH subsystem. The minus sign of the deviation rate indicates that the analytical value is less than the numerical one. Obviously, the deviation rates have a similar trend. Meanwhile, the peak power of the EMEH subsystem is the most affected. However, its magnitude is approximately one-fifth of that of the peak power generated from the piezoelectric subsystem. Therefore, linearized θ_{e} has almost the same effect on the peak power of the HEH and PEH subsystem.

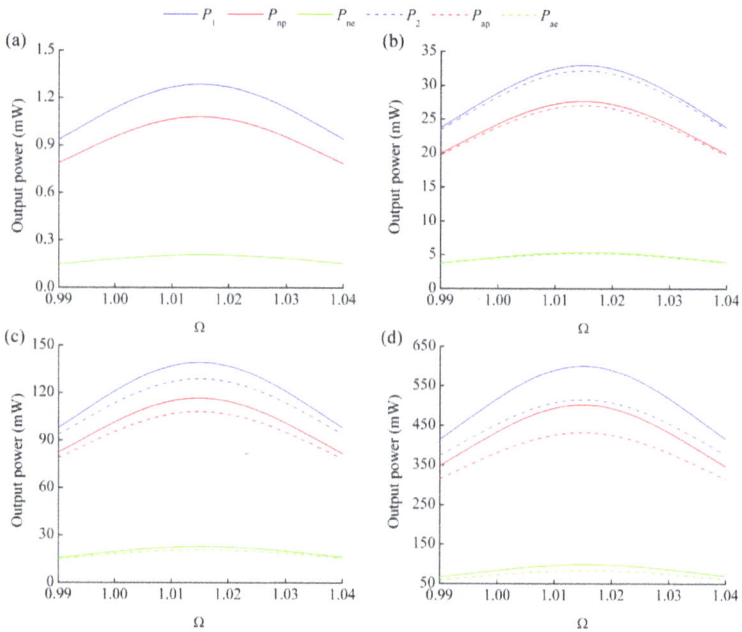

Figure 3. Output power versus excitation frequency ratio for different excitation accelerations: (**a**) 2 m/s²; (**b**) 10 m/s²; (**c**) 20 m/s²; (**d**) 40 m/s².

Table 2. Comparison of peak power from numerical and analytical solutions.

Acceleration (m/s²)	Numerical Results				Analytical Results			
	Ω_r	Peak Power (mW)			Ω_r	Peak Power (mW)		
		P_1	P_{np}	P_{ne}		P_2	P_{ap}	P_{ae}
2	1.015	1.288	1.081	0.207	1.015	1.287	1.080	0.207
10	1.015	32.969	27.635	5.334	1.015	32.175	27.002	5.173
20	1.015	139.291	116.588	22.703	1.015	128.702	108.008	20.694
40	1.015	600.311	502.117	98.194	1.015	514.806	432.031	82.775

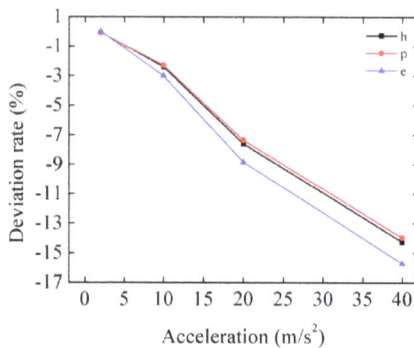

Figure 4. Deviation rates of the analytical values relative to numerical values. "h", "p" and "e" in the legend represent the HEH, the piezoelectric energy harvester (PEH) subsystem and the electromagnetic energy harvester (EMEH) subsystem.

3.2. Effects of Load Resistances on the Performance of the Hybrid Energy Harvester (HEH)

Based on the previous analysis, there is little difference between analytical and numerical solutions at low-level excitation acceleration. Consequently, the analytical solutions will be used in the later sections, and the excitation acceleration is set to 2 m/s².

As defined in the modeling section, λ_p and λ_e are the normalized electric resistances connected to the piezoelectric layers and induction coil, respectively. From Equation (12), the equivalent stiffness of the system can be expressed as:

$$K_{\text{eff}} = K + \frac{\theta_p^2 R_1^2 C_p \omega^2}{1 + R_1^2 C_p^2 \omega^2} + \frac{\theta_{\text{em}}^2 L_c \omega^2}{L_c^2 \omega^2 + (R_2 + R_c)^2} \tag{21}$$

It can be seen that load resistances can change the equivalent stiffness, as well as the resonant frequency.

Figure 5 shows the effects of normalized load resistances on the performance of the HEH. Obviously, λ_p plays a greater role in tuning the resonant frequency, as displayed in Figure 5a. With the increasing of λ_p, the resonant frequency gradually rises and tends to a stable value. However, the influence of λ_e seems to be negligible. Therefore, during the impedance matching of the experiment, the load resistance of the PEH subsystem should be preferentially determined, then that of the EMEH part. When $\lambda_p = 0$ and $\lambda_e = 0$, the piezoelectric layers are short-circuit, and the coil is open-circuit. At this time, $\Omega_r = 0.9996$, which denotes the normalized damped natural frequency and is named as the short-circuit resonant frequency sometimes [28]. When $\lambda_p = 6$, the piezoelectric layers are close to the open-circuit condition. The open-circuit resonant frequency is 1.0284 when the coil is open-circuit. This value is determined by the piezoelectric coupling term.

In Figure 5b, it can be seen that both load resistances affect the free-end displacement amplitude of the piezoelectric beam at the resonant frequency. With the increment of λ_p, this value first declines and then bounces back. However, it steadily decreases with the rising of λ_e. When $\lambda_p = 1.02$ and $\lambda_e = \lambda_c$, the amplitude reaches the lowest point 0.6189 mm, which is 42.86% the maximum value (1.4441 mm). It reveals that extracting electrical energy can suppress the vibration of the beam. Letting the excitation frequency ratio $\Omega = 1$, Figure 5c illustrates the output power of the HEH for different load resistances. It is clear that the output power first rises up and then falls off with the increasing of one load resistance, when the other one is kept constant. It reaches the maximum value 1.3730 mW when $\lambda_p = 0.55$ and $\lambda_e = 0.005$. According to the figure, we can conclude that hybrid energy harvesting provides an increment of output power, based on the parameters defined before. The energy conversion efficiency has the similar tendency as that of output power, as shown in Figure 5d. It reaches the highest 45.19% when $\lambda_p = 1$ and $\lambda_e = 0.012$. Comparison of Figure 5c,d shows that the matched load resistances for the optimal output power and energy conversion efficiency are not consistent. This is because the output power is related to the product of amplitude and efficiency. Furthermore, the optimal output power and energy conversion efficiency do not imply an optimal vibration suppression effect.

Figure 5e represents the varying of the operating frequency bandwidth. The bandwidth rises up with the increase of λ_e. However, it first increases and then declines when $\lambda_p = 1$. Due to the coupling of electromagnetic and piezoelectric energy harvesting, the electrical damping of the system increases, resulting in widening of the operating frequency bandwidth.

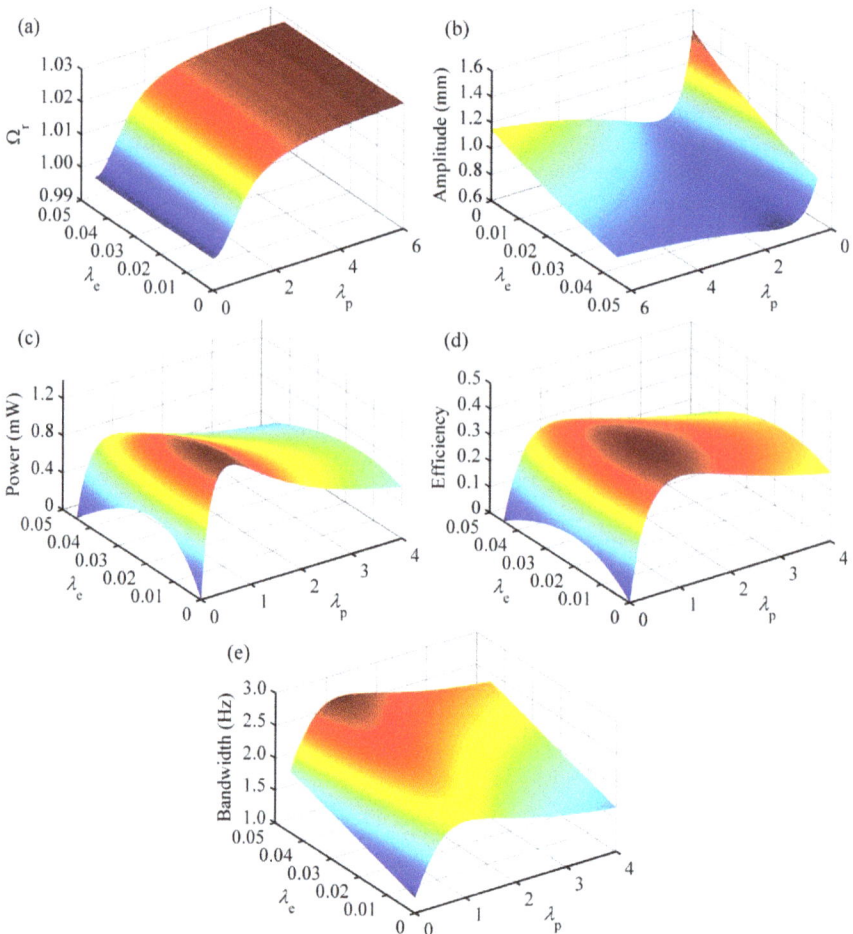

Figure 5. Performances of the HEH for different normalized load resistances λ_p and λ_e: (**a**) normalized resonant frequency Ω_r; (**b**) free-end displacement amplitude of the piezoelectric beam at the resonant frequency; (**c**) output power; (**d**) energy conversion efficiency; (**e**) operating frequency bandwidth.

3.3. Effects of Electromechanical Coupling Factors on the Performance of the HEH

In previous literature [29,30], scholars took κ_p^2/ζ_m as the indicator of the piezoelectric coupling effect. In this paper, we define the piezoelectric coupling factor $\sigma_p = \kappa_p^2/\zeta_m$. Similarly, we take electromagnetic coupling factor $\sigma_e = \kappa_e^2/\zeta_m$ as the indicator of the electromagnetic coupling effect. Based on the parameters defined before, the values of σ_p and σ_e are about 3.0 and 25.0, respectively. Ω is still set to one.

Figure 6 plots the optimal output power and energy conversion efficiency of the HEH with matched load resistances for different σ_p and σ_e. The power climbs sharply at first and then slows down as σ_p increases. There is also a growing tendency of the optimal power with the increase of σ_e, although the influence of σ_e almost fades off when σ_p is larger than eight. Apparently, σ_p exerts stronger influence on the power than σ_e. In a word, when the piezoelectric coupling effect is weak or medium, HEH generates more power than the single-mechanism energy harvester. With the enhancement of coupling effects, the efficiency also steadily increases. According to the figure, we can conclude that hybrid energy harvesting mechanism contributes to improving the energy conversion efficiency of VEH.

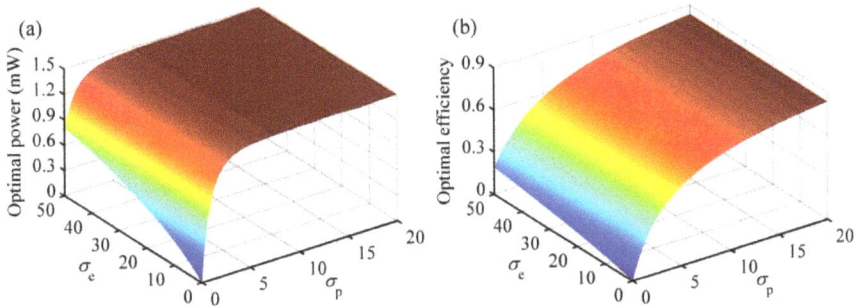

Figure 6. Optimal output power and energy conversion efficiency for different coupling factors σ_p and σ_e: (**a**) output power; (**b**) energy conversion efficiency.

3.4. Effect of Mechanical Damping Ratio on the Performance of the HEH

Mechanical damping can consume the energy of the HEH and limits the generating performance. To observe the effect of mechanical damping, Figure 7 provides the output power increment of the HEH relative to the conventional PEH with matched load resistances for different Ω and ζ_m. Both matched values of λ_p connected to HEH and PEH are 0.55. The matched λ_e is 0.005. It can be seen from the figure that when $\zeta_m < 0.041$, there are two peaks in the power increment, corresponding to the vicinity of excitation frequency ratios of 0.98 and 1.03. On the contrary, there is only one peak around $\Omega = 1.03$. The appearance of two peaks is due to the broader bandwidth of the HEH. In addition, the minimum power increment dramatically reduces to the negative value with the decrease of ζ_m. It implies that the peak output power of the PEH is larger than that of the HEH under the condition of very small mechanical damping. When ζ_m is bigger than 0.015, the generating performance of the HEH is definitely better than that of the PEH. Overall, the mechanical damping ratio can affect the superiority of the HEH to the PEH.

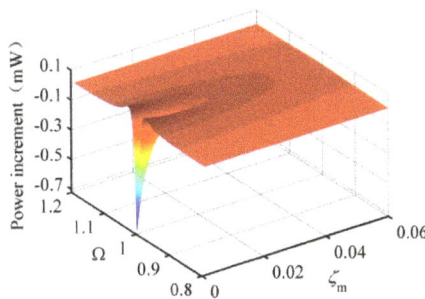

Figure 7. The output power increment of the HEH relative to the conventional PEH with matched load resistances for different Ω and ζ_m.

3.5. Effects of Coil Parameters on the Performance of the HEH

In order to evaluate the effects of the induction coil parameters on the generating performance of the HEH, we analyzed the effects of height h_c and outer radius R_o of the coil on the inductive reactance $|Z_L|$, resonance frequency ω_r, electromagnetic coupling coefficient θ_e and electromagnetic coupling factor σ_e, as shown in Figure 8.

Note that $|Z_L| = \omega L_c$. Considering the frequencies of most vibration sources in the environment below 200 Hz [31], the excitation frequency in the analytical model is assumed to be 200 Hz. For a certain geometry of the coil, the copper wire is supposed to be tightly wound and entirely fill the coil volume. The external load resistance R_2 is equal to R_c. When h_c and R_o increase, the coil turns and R_c will increase, as well as the ratio of $|Z_L|$ to R_c. As designed in this paper, R_o is 13.15 mm, and h_c is 17 mm. The corresponding ratio is about 0.35, which is much higher than the actual value. If the matched load resistance R_2 is taken into account, the ratio of $|Z_L|$ to $(R_c + R_2)$ will be close to 0.18. In an integral energy harvesting system, the proportion of $|Z_L|$ relative to the total impedance of the system will further drop due to the impedance introduced by the external circuit. In Figure 8b, the load resistance R_1 is set zero to exclude the influence of the piezoelectric coupling term on the resonant frequency. As can be seen, the ratio reaches the maximum value 1.05 when $h_c = 21.1$ mm and $R_o = 21.4$ mm. Hence, the resonant frequency of the vibration energy harvester can be tuned by changing the parameters of the induction coil. However, the frequency tuning range is so limited that many scholars neglected the effect of inductive reactance on the resonant frequency for the simplification.

There is a notable distinction between electromagnetic coupling coefficient θ_e and coupling factor σ_e for different h_c and R_o. θ_e gradually increases with the increase of h_c or R_o. As reported before [11], the inductive electromotive force is proportional to θ_e. However, this does not mean that the higher θ_e is, the better the generating performance of the HEH is. σ_e shows a downward trend with the increase of R_o and reaches the peak when $h_c = 13.1$ mm and $R_o = 12.3$ mm. This is mainly because the increase of h_c and R_o induces the increase of inductance of the coil. Referring to Figure 6, the higher the value of σ_e is, the more electric power that can be generated. Therefore, the coil parameters of the HEH should be optimized according to the variation of σ_e.

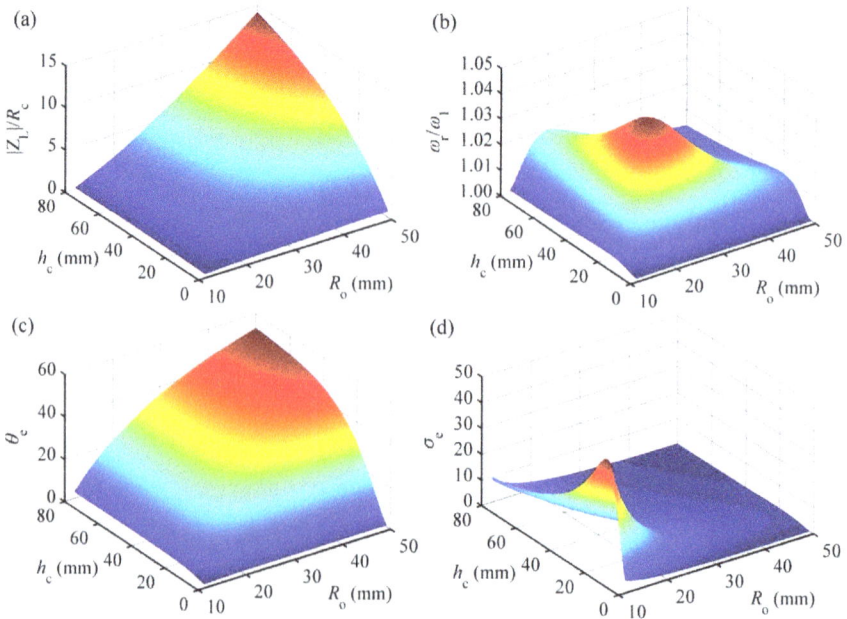

Figure 8. Performance of the HEH for different h_c and R_o: (**a**) ratio of inductive reactance $|Z_L|$ to internal resistance R_c of the coil; (**b**) ratio of resonant frequency ω_r to undamped natural frequency ω_1; (**c**) electromagnetic coupling coefficient θ_e; (**d**) electromagnetic coupling factor σ_e.

3.6. Scaling Effects on the Performance of the HEH

To observe the size-scale effect on the performance, we hold all parameters constant except the geometric dimensions. Define α as the scaling factor. As a sample, all geometric dimensions of the HEH are scaled down by 100 times when $\alpha = 0.01$. Five cases with scaling factors of 1, 0.1, 0.01, 0.001 and 0.0001 are calculated to explore the scaling effects. That is to say, the performance of the HEH is analyzed with the size range from the mm to the μm scale. Figure 9 displays the mechanical and electrical performance of the HEH for different scaling factors. Figure 9a shows a positive correlation between the effective stiffness and scaling factor. Therefore, the beam can deform more and more under the same transverse load as the size is scaled down. However, the undamped resonant frequency f_1 of the piezoelectric beam increases with the size reduction. That is due to greater shrinking of the mass by α^3. Piezoelectric and electromagnetic coupling factors have no size dependence under the assumption that the material parameters remain unchanged. As a resonant VEH, it is clear that only in the vicinity of the resonant frequency can the HEH have excellent generating performance. Therefore, the inductive reactance $|Z_L|$ at undamped resonant frequency and its effect on the resonant frequency ω_r are analyzed, as illustrated in Figure 9c. Because of the increment of undamped resonant frequency, the ratio of $|Z_L|$ to R_c gradually increases with the size scaling down. Although inductive reactance is enhanced, the ratio of ω_r to ω_1 rises first, then levels off when α is less than 0.001. We can find the reason from Equation (22):

$$\frac{\omega_r}{\omega_1} = \sqrt{1 + \kappa_e^2 \cdot \frac{Z_L^2}{Z_L^2 + (R_2 + R_c)^2}} \tag{22}$$

When α is less than 0.001, $|Z_L|$ is much larger than R_c. Generally, R_2 and R_c have the same order of magnitude. Consequently, this ratio gradually approaches $\sqrt{1 + \kappa_e^2}$, which is not affected by the size scaling. The normalized load resistance λ_p keeps constant, while λ_e increases first, then levels off when α is less than 0.01.

To compare the output power with different size scale, we investigate the power density (PD) of the HEH for different scaling factors and excitation frequency ratios Ω. As shown in Figure 9e, PD is proportional to the scaling factor. Therefore, it is a challenging issue to improve the PD of MEMS-scale VEH. Figure 9f shows the maximum output power of the HEH (P_{hmax}), stand-alone PEH (P_{spmax}) and stand-alone EMEH (P_{semax}) with matched load resistances and resonant frequency ratios Ω_r for different α. As the induction coil is open-circuit and only R_1 is connected, the HEH is changed into a stand-alone PEH. When R_2 is connected to the induction coil and the piezoelectric layers are directly connected without external load resistance (short-circuit condition), the stand-alone EMEH is developed. Firstly, there is almost the same trend for the resonant frequencies of the HEH and stand-alone EMEH, which increase with the size scaling down and then remain stable. The resonant frequency of the stand-alone PEH is not affected. All maximum output power drops dramatically as the size decreases. Note that the value of P_{hmax} is greater than that of P_{spmax} and P_{semax} for any α, based on the geometric parameters listed before. It demonstrates again that hybrid energy harvesting can enhance the output power of the VEH even at the MEMS scale.

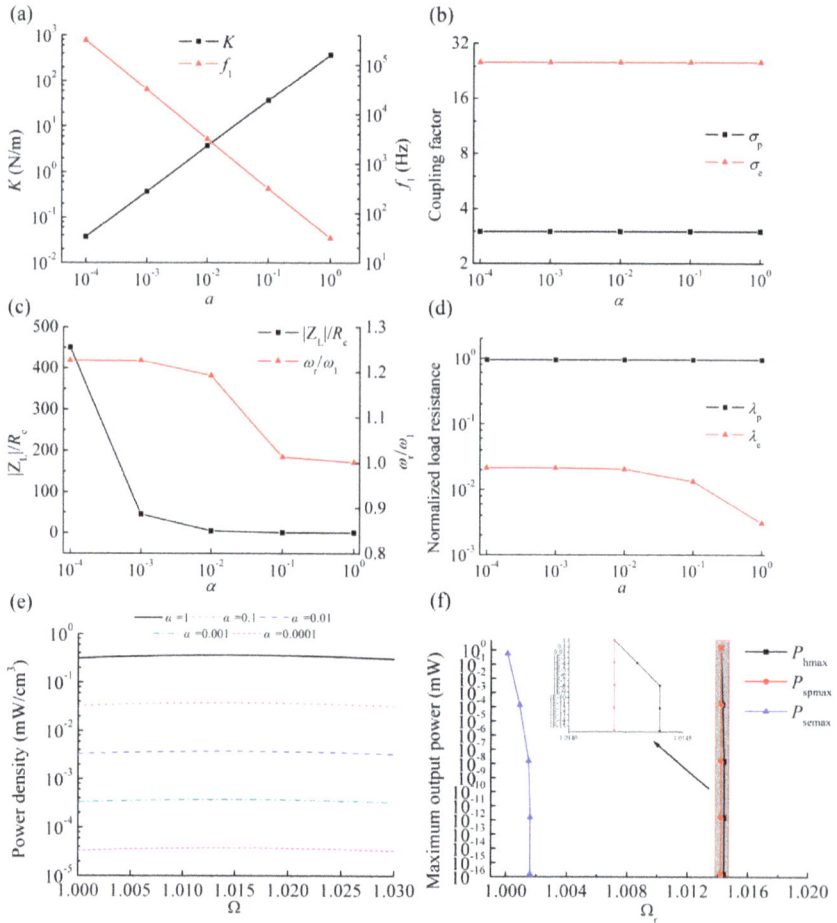

Figure 9. Performance of the HEH for different scaling factor α: (**a**) effective stiffness and undamped resonant frequency; (**b**) coupling factor ratio; (**c**) ratios of $|Z_L|/R_c$ and ω_r/ω_1; (**d**) normalized load resistances; (**e**) power density; (**f**) Maximum output power and resonant frequency ratios for HEH, stand-alone PEH and stand-alone EMEH.

4. Experiment

In order to verify the established theoretical model, an experimental test system was assembled, as shown in Figure 10. A sinusoidal-wave excitation signal is generated by a signal generator (DG-1022, Rigol Technologies Inc., Beijing, China), amplified by a power amplifier (YE5874A, Sinocera Piezotronics Inc., Yangzhou, China) and used to control the vibration of an electromagnetic shaker (JZK-50, Sinocera Piezotronics Inc., Yangzhou, China). The vibration frequency and acceleration are acquired by an accelerometer (YD64-310, Qinhuangdao Xinheng Electronic Technology Co., Ltd., Qinhuangdao, China) and conditioned by a charge amplifier (CA-3, Qinhuangdao Xinheng Electronic Technology Co. Ltd., Qinhuangdao, China). The generated voltage and acceleration signal are input into a computer through the data acquisition module (NI 9229, National Instruments Co., Austin, TX, USA).

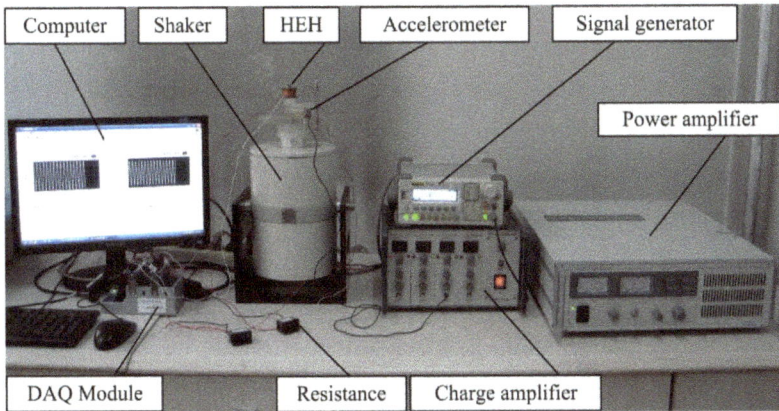

Figure 10. Experimental setup for the HEH.

Based on the previous analysis, the output power of the HEH is optimal when the load resistances connected are matched. In order to compare the output power frequency responses of the stand-alone PEH, stand-alone EMEH and HEH, it is necessary to respectively match the load resistances. Figure 11 shows the output power of the stand-alone PEH for different R_1 at the resonant frequency. The base excitation was harmonic, and acceleration was kept at 2 m/s². Through the experimental test, the output power achieved the maximum 0.887 mW when the matched resistance R_1 was 52 kΩ. When R_1 was reduced to 1 kΩ, the piezoelectric layers were considered to be short-circuit connected, and the piezoelectric damping was negligible. Under this condition, the mechanical damping ratio of the piezoelectric beam was measured to be 0.0295 by using the logarithmic decrement method.

Figure 12 represents the output power of the stand-alone EMEH for different R_2 at the resonant frequency. The output power reaches the peak 0.377 mW when R_2 was equal to 480 Ω. The measured internal resistance of the coil was 338 Ω. Letting $R_2 = 10$ kΩ produced an approximately open-circuit condition for the coil. The mechanical damping ratio of the magnetic oscillator was measured to be 0.025. Compared with that of the piezoelectric oscillator, the change of mechanical damping ratio is mainly due to the secondary piezoelectric effect of the piezoelectric material, which induces more damping to the system.

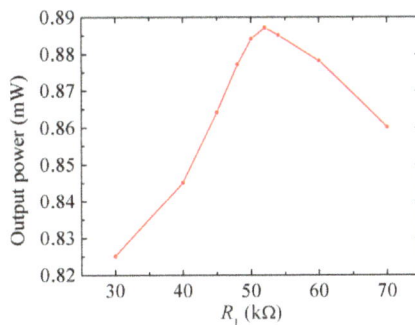

Figure 11. The output power of the stand-alone PEH for different R_1 at the resonant frequency.

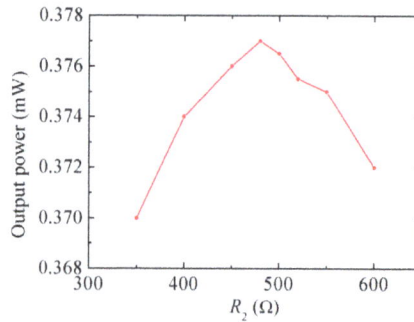

Figure 12. The output power of the stand-alone EMEH for different R_2 at the resonant frequency.

Figure 13 illustrates the output power of the HEH for different R_2 at the resonant frequency, when the piezoelectric layers were connected with the matched R_1. The maximum power 0.913 mW occurs when $R_2 = 1850\ \Omega$ at the resonant frequency. It can be seen that the matched R_2 is different from that of the stand-alone EMEH. This is because the resonant frequency of the beam increases when matched R_1 is connected to the piezoelectric layers, which results in the change of the matched R_2.

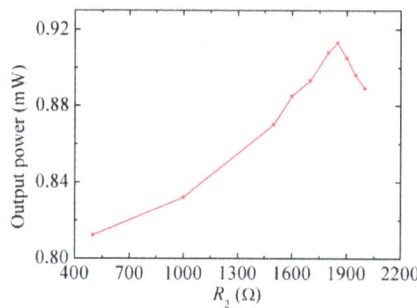

Figure 13. The output power of the HEH for different R_2 at the resonant frequency.

At last, the optimal output power of the stand-alone PEH, stand-alone EMEH and HEH at different excitation frequencies was tested in turn, as shown in Figure 14. Note that the subscripts "esp", "ese" and "eh" represent the experimental results of the stand-alone PEH, stand-alone EMEH, and HEH, respectively. "asp" and "ase" stand for the analytical results of the stand-alone PEH and stand-alone EMEH, respectively. "nsp" and "nse" mean the numerical results of the stand-alone PEH and stand-alone EMEH, respectively. The load resistances corresponding to the optimal output power were the matched ones. It can be seen that the numerical results nearly coincide with the analytical results. The experimental results also show good agreement with the theoretical values. It proves that the theoretical model is valid. The measured peak output power of the HEH (0.913 mW) is 2.93% and 142.18% higher than that of the stand-alone PEH (0.887 mW) and EMEH (0.377 mW), respectively. Moreover, the operation frequency bandwidth of the HEH is the widest, which (about 2.7 Hz) increases up to 108%- and 122.7%-times that of stand-alone PEH (about 2.5 Hz) and EMEH (about 2.2 Hz), respectively. Note that σ_p and σ_e of the HEH prototype are 2.03 and 17.44, respectively. Consequently, the superiority of the hybrid energy harvesting mechanism is in accordance with the previous theoretical analysis. The measured resonant frequency of the HEH is 31.2 Hz, which is the same as that of the stand-alone PEH, but higher than that of the stand-alone EMEH (30.6 Hz). It indicates that coupling electromagnetic energy harvesting has little effect on the system resonant frequency. In addition, the tested resonant frequencies are lower than the theoretical values, which is mainly caused by the softened spring effect of the resonant structure [32].

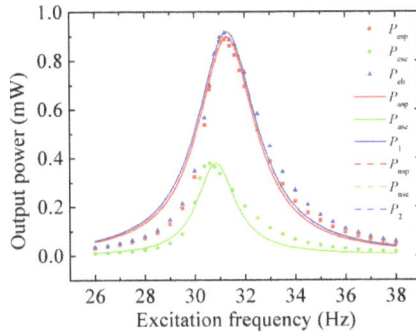

Figure 14. The optimal output power of the stand-alone PEH, stand-alone EMEH and HEH at different excitation frequencies.

5. Conclusions

In order to evaluate the generating performance of the piezoelectric-electromagnetic hybrid energy harvester, this paper developed an approximate distributed-parameter theoretical model of the HEH based on the energy method and Euler–Bernoulli beam theory. The analytical solutions were compared with the numerical solutions and used to observe the influences of mechanical and electric parameters on the generating characteristics and dynamic responses of the HEH. A meso-scale HEH prototype was fabricated and tested. The experimental results verified the theoretical model and analysis. The following conclusions were obtained:

The difference between numerical and analytical solutions gradually increases with the increment of excitation acceleration. The load resistance connected to piezoelectric layers has a significant effect on tuning the resonant frequency of the HEH, while the effect of that connected to the coil can be neglected. Therefore, the load resistance of the piezoelectric energy harvesting subsystem should be preferentially determined during the impedance matching of the experiment. The matched load resistances for the optimal output power and energy conversion efficiency are not consistent. Extracting electrical energy can suppress the vibration of beam. However, the optimal output power and energy conversion efficiency do not imply an optimal vibration suppression effect. Regardless of the piezoelectric coupling strength, coupled electromagnetic and piezoelectric energy harvesting results in widening operating frequency bandwidth and improving the energy conversion efficiency. However, the HEH generates more power than the single-mechanism energy harvester only when the piezoelectric coupling effect is weak or medium. The piezoelectric coupling factor of the HEH prototype is 2.03, which denotes the medium coupling. Its maximum output power (0.913 mW) with matched load resistances is 2.93% and 142.18% higher than that of the stand-alone PEH (0.887 mW) and EMEH (0.377 mW), respectively. The operation frequency bandwidth of the HEH is 108%- and 122.7%-times of that of stand-alone PEH and EMEH, respectively. The superiority of the HEH on the output power to the stand-alone PEH is also affected by the mechanical damping ratio. The influence of the inductive coil on the system resonant frequency can be neglected when the frequency of the vibration source is less than 200 Hz. For the optimal output power, the coil parameters of the HEH should be optimized according to the electromagnetic coupling factor. When the size is scaled down to the micro scale, some mechanical and electrical performance is affected. However, it indicates that hybrid energy harvesting can enhance the output power of the VEH even at the MEMS scale.

The numerical model can be used for the optimization of the HEH to harvest maximum power from a given excitation source. However, the accuracy of the analytical model can be guaranteed at the low-level excitation acceleration. Furthermore, it has a better computational efficiency. The superiority of the HEH relative to the stand-alone PEH or EMEH depends on mechanical and electric parameters. In future work, an integrated system including HEH and interface circuit will be designed, analyzed and optimized.

Acknowledgments: The work described in this paper was supported by the National Natural Science Foundation of China (Grant No. 51677043).

Author Contributions: All authors conceived of and designed the experiments. Zhenlong Xu and Xiaobiao Shan set up the experimental platform. Zhenlong Xu and Hong Yang conducted the experiments. All authors contributed to analyzing the experimental data and writing the paper.

Conflicts of Interest: The authors declare no conflict of interest.

References

1. Roundy, S.; Wright, P.K. A piezoelectric vibration based generator for wireless electronics. *Smart Mater. Struct.* **2004**, *13*, 1131–1142. [CrossRef]
2. Glynne-Jones, P.; Tudor, M.J.; Beeby, S.P.; White, N.M. An electromagnetic, vibration-powered generator for intelligent sensor systems. *Sens. Actuators A* **2004**, *110*, 344–349. [CrossRef]
3. Le, C.P.; Halvorsen, E. MEMS electrostatic energy harvesters with end-stop effects. *J. Micromech. Microeng.* **2012**, *22*, 074013. [CrossRef]
4. Mori, K.; Horibe, T.; Ishikawa, S.; Shindo, Y.; Narita, F. Characteristics of vibration energy harvesting using giant magnetostrictive cantilevers with resonant tuning. *Smart Mater. Struct.* **2015**, *24*, 125032. [CrossRef]
5. Khaligh, A.; Zeng, P.; Wu, X.; Xu, Y. A Hybrid Energy Scavenging Topology for Human-Powered Mobile Electronics. In Proceedings of the 34th Annual Conference of the IEEE Industrial Electronics Society, Orlando, FL, USA, 10–13 November 2008; pp. 448–453.
6. Challa, V.R.; Prasad, M.G.; Fisher, F.T. A coupled piezoelectric-electromagnetic energy harvesting technique for achieving increased power output through damping matching. *Smart Mater. Struct.* **2009**, *18*, 095029. [CrossRef]
7. Yang, B.; Lee, C.; Kee, W.L.; Lim, S.P. Hybrid energy harvester based on piezoelectric and electromagnetic mechanisms. *J. Micro/Nanolith. MEMS MOEMS* **2010**, *9*, 023002.023001–023002.023010. [CrossRef]
8. Tadesse, Y.; Zhang, S.; Priya, S. Multimodal energy harvesting system: Piezoelectric and electromagnetic. *J. Intell. Mater. Syst. Struct.* **2009**, *20*, 625–632. [CrossRef]
9. Shan, X.-B.; Guan, S.-W.; Liu, Z.-S.; Xu, Z.-L.; Xie, T. A new energy harvester using a piezoelectric and suspension electromagnetic mechanism. *J. Zhejiang Univ.-Sci. A* **2013**, *14*, 890–897. [CrossRef]
10. Wang, H.-Y.; Tang, L.-H.; Guo, Y.; Shan, X.-b.; Xie, T. A 2DOF hybrid energy harvester based on combined piezoelectric and electromagnetic conversion mechanisms. *J. Zhejiang Univ.-Sci. A* **2014**, *15*, 711–722. [CrossRef]
11. Xu, Z.; Shan, X.; Chen, D.; Xie, T. A novel tunable multi-frequency hybrid vibration energy harvester using piezoelectric and electromagnetic conversion mechanisms. *Appl. Sci.* **2016**, *6*, 10. [CrossRef]
12. Karami, M.A.; Inman, D.J. Equivalent damping and frequency change for linear and nonlinear hybrid vibrational energy harvesting systems. *J. Sound Vib.* **2011**, *330*, 5583–5597. [CrossRef]
13. Li, P.; Gao, S.; Cai, H.; Wu, L. Theoretical analysis and experimental study for nonlinear hybrid piezoelectric and electromagnetic energy harvester. *Microsyst. Technol.* **2015**, *22*, 727–739. [CrossRef]
14. Halim, M.A.; Cho, H.O.; Park, J.Y. A handy-motion driven, frequency up-converted hybrid vibration energy harvester using PZT bimorph and nonmagnetic ball. *J. Phys.* **2014**, *557*, 012042. [CrossRef]
15. Edwards, B.; Hu, P.A.; Aw, K.C. Validation of a hybrid electromagnetic–piezoelectric vibration energy harvester. *Smart Mater. Struct.* **2016**, *25*, 055019. [CrossRef]
16. Deng, L.; Wen, Z.; Zhao, X. Theoretical and experimental studies on piezoelectric-electromagnetic hybrid vibration energy harvester. *Microsyst. Technol.* **2016**, *23*, 935–943. [CrossRef]
17. Xia, H.; Chen, R.; Ren, L. Analysis of piezoelectric–electromagnetic hybrid vibration energy harvester under different electrical boundary conditions. *Sens. Actuators A* **2015**, *234*, 87–98. [CrossRef]
18. Li, P.; Gao, S.; Niu, S.; Liu, H.; Cai, H. An analysis of the coupling effect for a hybrid piezoelectric and electromagnetic energy harvester. *Smart Mater. Struct.* **2014**, *23*, 065016. [CrossRef]

19. Shan, X.; Xu, Z.; Song, R.; Xie, T. A new mathematical model for a piezoelectric-electromagnetic hybrid energy harvester. *Ferroelectrics* **2013**, *450*, 57–65. [CrossRef]

20. Mahmoudi, S.; Kacem, N.; Bouhaddi, N. Enhancement of the performance of a hybrid nonlinear vibration energy harvester based on piezoelectric and electromagnetic transductions. *Smart Mater. Struct.* **2014**, *23*, 075024. [CrossRef]

21. Sodano, H.A.; Park, G.; Inman, D.J. Estimation of electric charge output for piezoelectric energy harvesting. *Strain* **2004**, *40*, 49–58. [CrossRef]

22. Dutoit, N.E.; Wardle, B.L.; Kim, S.-G. Design considerations for MEMS-scale piezoelectric mechanical vibration energy harvesters. *Integr. Ferroelectr.* **2005**, *71*, 121–160. [CrossRef]

23. Inman, D.J.; Singh, R.C. *Engineering Vibration*, 3rd ed.; Pearson Education: Upper Saddle River, NJ, USA, 2008.

24. Meitzler, A.; Tiersten, H.F.; Warner, A.W.; Berlincourt, D.; Couqin, G.A.; Welsh, F.S., III. *IEEE Standard on Piezoelectricity*; American National Standards Institute: Washington, DC, USA, 1976.

25. Harne, R.L. Theoretical investigations of energy harvesting efficiency from structural vibrations using piezoelectric and electromagnetic oscillators. *J. Acoust. Soc. Am.* **2012**, *132*, 162–172. [CrossRef] [PubMed]

26. Zeng, P.; Khaligh, A. A permanent-magnet linear motion driven kinetic energy harvester. *IEEE Trans. Ind. Electron.* **2013**, *60*, 5737–5746. [CrossRef]

27. Cammarano, A.; Neild, S.A.; Burrow, S.G.; Inman, D.J. The bandwidth of optimized nonlinear vibration-based energy harvesters. *Smart Mater. Struct.* **2014**, *23*, 055019. [CrossRef]

28. Erturk, A.; Inman, D.J. An experimentally validated bimorph cantilever model for piezoelectric energy harvesting from base excitations. *Smart Mater. Struct.* **2009**, *18*, 025009. [CrossRef]

29. Tang, L.; Yang, Y. Analysis of synchronized charge extraction for piezoelectric energy harvesting. *Smart Mater. Struct.* **2011**, *20*, 085022. [CrossRef]

30. Shu, Y.C.; Lien, I.C. Analysis of power output for piezoelectric energy harvesting systems. *Smart Mater. Struct.* **2006**, *15*, 1499–1512. [CrossRef]

31. Roundy, S.; Wright, P.K.; Rabaey, J. A study of low level vibrations as a power source for wireless sensor nodes. *Comput. Commun.* **2003**, *26*, 1131–1144. [CrossRef]

32. Bouendeu, E.; Greiner, A.; Smith, P.J.; Korvink, J.G. A low-cost electromagnetic generator for vibration energy harvesting. *IEEE Sens. J.* **2011**, *11*, 107–113. [CrossRef]

MDPI

St. Alban-Anlage 66

4052 Basel

Switzerland

Tel. +41 61 683 77 34

Fax +41 61 302 89 18

www.mdpi.com

Micromachines Editorial Office

E-mail: micromachines@mdpi.com

www.mdpi.com/journal/micromachines

www.ingramcontent.com/pod-product-compliance
Lightning Source LLC
Chambersburg PA
CBHW051859210326
41597CB00033B/5955